I0064255

An Unbounded Experience in Random Walks with Applications

An Unbounded Experience in Random Walks with Applications

Michael F Shlesinger

World Scientific

NEW JERSEY · LONDON · SINGAPORE · BEIJING · SHANGHAI · HONG KONG · TAIPEI · CHENNAI · TOKYO

Published by

World Scientific Publishing Co. Pte. Ltd.

5 Toh Tuck Link, Singapore 596224

USA office: 27 Warren Street, Suite 401-402, Hackensack, NJ 07601

UK office: 57 Shelton Street, Covent Garden, London WC2H 9HE

Library of Congress Cataloging-in-Publication Data
Names: Shlesinger, Michael F., author.
Title: An unbounded experience in random walks with applications / Michael F. Shlesinger.
Description: New Jersey : World Scientific, [2021] | Includes index.
Identifiers: LCCN 2020055878 | ISBN 9789811232800 (hardcover) |
    ISBN 9789811232817 (ebook for institutions) | ISBN 9789811232824 (ebook for individuals)
Subjects: LCSH: Random walks (Mathematics)
Classification: LCC QA274.73 .S55 2021 | DDC 519.2/82--dc23
LC record available at https://lccn.loc.gov/2020055878

British Library Cataloguing-in-Publication Data
A catalogue record for this book is available from the British Library.

Copyright © 2021 by World Scientific Publishing Co. Pte. Ltd.

All rights reserved. This book, or parts thereof, may not be reproduced in any form or by any means, electronic or mechanical, including photocopying, recording or any information storage and retrieval system now known or to be invented, without written permission from the publisher.

For photocopying of material in this volume, please pay a copying fee through the Copyright Clearance Center, Inc., 222 Rosewood Drive, Danvers, MA 01923, USA. In this case permission to photocopy is not required from the publisher.

For any available supplementary material, please visit
https://www.worldscientific.com/worldscibooks/10.1142/12176#t=suppl

Desk Editors: Vishnu Mohan/Lai Fun Kwong

Typeset by Stallion Press
Email: enquiries@stallionpress.com

Dedicated to
Nava, Adi, Josh, Jordan, Leo, Jacob, Ben, Sara, Ellie,
Natalie, Dan, Michelle, Emma

Typing this book at home, September 2020.

Preface

In September of 1970, I was starting graduate school in physics and looking for a thesis advisor. I knocked on the door of Elliott Waters Montroll, the Einstein Professor of Physics at the University of Rochester. This book is about what happened next. This book is meant to interest students in the area of stochastic processes and to show that one need not be famous to have a great career and life in physics. I have kept the book short enough in the hope that it will be read.

This is certainly not a text book, so I won't cite a long list of papers, but instead point the reader to a number of texts on probability. I won't go deeply into physics, but instead the focus will be on math, the type of applied math that is connected to physics. I narrowly focus on the topic of random walks, starting with walks on a lattice and then generalizations to include scale invariant jump processes, internal states, coupled space and time memories and applications in condensed matter physics. Back in 1979, I wrote a paper of nearly 100 pages just on random walks with internal states, so it's easy to fill the pages of a long heavy tome. Here, I try to give the flavor and a taste, not the whole meal.

The book is about stochastic processes, that is those that change with time. So, topics like the probability to draw three aces in a poker hand, or eigenvalues of random matrices won't be found here. I leave out vast topics of stochastic processes including Langevin and Fokker Planck equations and their fractional variants. Also, the topics of random walks on fractals, noise, and equations of motion with noisy parameters are not the concern here. But there is enough to write a book about just with considering some random walks. If you like the random walk topic, then you can move on to the array of other topics involving probability theory. At a meeting in Lanzhou, China, a student asked for my opinion on the next great thing in probability. My answer is, the next advance is over the horizon, and what appears will only be limited by one's imagination. Maybe that student will find the answer.

I include some of my experiences and observations as a student and later as a physicist (but I do not cover my 37 plus years at the Office of Naval Research, managing a nonlinear physics program, which has its own rich collection of anecdotes, perhaps for future story telling). I hope this book is readable enough and entertaining enough that it will be read, and it was personally fun enough for me to write and to pay homage to my mentors Elliott Waters Montroll and Harvey Scher.

Despite the lacuna in my exposition, there are excellent choices for the reader who desires more complete and more formal works. I recommend:

Random Walks and Random Environments (Oxford Science Publications, 1995) by Barry D. Hughes. When I reviewed this book, I wrote that Hughes has written a classic and the field of random walks finally has a book worthy of its accomplishments.

This book should appeal both to mathematicians and scientists. It is a veritable encyclopedia of random walks and percolation. The work comes in two volumes.

Aspects and Applications of the Random Walk (North-Holland, 1994) by George H. Weiss. A wonderful exposition of random walks, by an expert, in a style for physicists. It has a number of sophisticated mathematical techniques and tricks that the practitioner will find useful.

First Steps in Random Walks: From Tools to Application (Oxford University Press, 2011) by J. Klafter and I. M. Sokolov. Includes a modern discussion of fractional diffusion equations.

Power Laws: A Statistical Trek (Springer, 2020) by Iddo Eliazar. Deft use of the power Poisson process to capture the behavior of a wide variety of phenomena, including extreme value statistics.

All of the above are friends and co-authors of mine. I also recommend:

A Guide to First Passage Processes (Cambridge University Press, 2007) by Sid Redner. Covers a variety of first passage topics, geometries and situations, such as reactions as a first passage time problem and properties of fractal networks.

For those beginning with probability, William Feller's, *An Introduction to Probability*, Vol.1, is a great start. Volume 2 is a more difficult and more rambling.

A more recent and excellent work is Papoulis and Pillai, *Probability, Random Variables, and Stochastic Processes*, 4th edition (McGraw-Hill, 2002).

Back in 1865, Isaac Todhunter, wrote *The History of Probability Theory*. His treatise focused on the works of Laplace (who himself borrowed from earlier works), and earlier books of Huygens, Jacob Bernoulli, Montmort, and DeMoivre.

For the early history of probability there is the delightful book:

Gods, Games, and Gambling: A History of Probability and Statistical Ideas (Charles Griffin, London, 1962; Dover reprint 1998) by Florence Nightingale David. Covers ideas from the earliest times up to the Newtonian Era. She was named after Florence Nightingale, a family friend.

Michael F. Shlesinger
Rockville, MD, USA

About the Author

Dr Michael F. Shlesinger has worked in universities, a non-profit research institute, and for the last 37 years in the Office of Naval Research. There he has headed the physics division and worked as a chief scientist and as a member of the senior executive service with the presidential rank award and the Saalfeld outstanding lifetime achievement award. He held the Kinnear professorship at the USNA. He is a fellow of the American Physical Society with their outstanding referee award and has published works on stochastic processes including cover articles in *Physics Today*. In another field he has patents on protein design. He co-founded the World Scientific journal *Fractals* and was a divisional associate editor of *Physical Review Letters*. He co-founded the Experimental Chaos Conference. He has given invited talks including the Michelson Lecture at the USNA and the Regents Lecture at UCSD.

The Zaslavsky Map

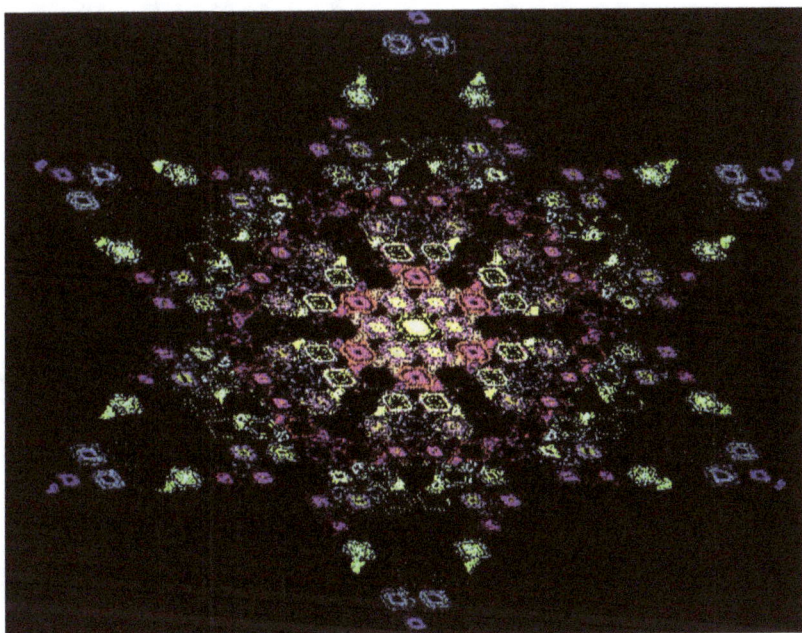

The evolution of an ensemble of trajectories (color coded by time) starting near the origin of the six-fold symmetric Zaslavsky map. The individual trajectories resemble a random walk, to be discussed latter in terms of a Lévy flight.

Contents

CHAPTER 1

Random Walks on a Lattice

1.1. Probabilities

Let's jump right in and see if by the end of the first chapter you want to continue and delve more deeply into random walks. We start simply with a 1D lattice of equally spaced sites a distance of unity apart and a random walker starting at the origin which is labeled as 0. Now let the walker jump one lattice site at a time to the right or left, each with probability 1/2. So, after one jump the walker is either at +1 or −1. After two jumps the walker is at +2 or −2 each with probability 1/4, or back at the origin with probability of 1/2. Playing with combinatorics one can find that the well-known probability $P_n(l)$ of being at site l after n jumps is

$$P_n(l) = \frac{n!}{[\frac{1}{2}(n+l)!][\frac{1}{2}(n-l)!]} 2^{-n} \tag{1.1}$$

Staring at this expression it is not obvious of what this looks like for large n. We could use Stirling's formula for $n!$ for large n and show that this converges to a Gaussian. That's exactly what DeMoivre did way back in 1756, although it was not called a Gaussian as Gauss did not publish on the "Gaussian" until 1809. We won't derive that result because it won't solve the probability of what if we allow jumps of all integer sizes and of unequal probabilities of jumping to the right and left. Also, what if we allow jumps on a 2D, 3D, or higher dimensional lattice?

Let's seemingly complicate the approach by going to complex numbers. Look at the expression,

$$\left(\frac{1}{2}\exp(i\,l) + \frac{1}{2}\exp(-i\,l)\right)^n$$

and calculate the coefficient of the exponentials as a function of n. For $n = 2$, we have

$$\frac{1}{4}\exp(2i\,l) + \frac{1}{2}\exp(0) + \frac{1}{4}\exp(-2i\,l)$$

and we can see that the coefficients match the probabilities for being at 2, 0, -2. This works for symmetric nearest-neighbor jumps, but what if we allow any jump length and probability for that jump.

Let's start again differently with a general equation and jump to complex numbers through use of the Fourier transform. Start with,

$$P_{n+1}(l) = \sum_{l'} P_n(l - l')P_1(l') \tag{1.2}$$

which accounts for being at l after $n + 1$ jumps by being at $l - l'$ at the nth jump and then having the final jump being of displacement l'. Our l's can be lengths in 1D or they can be vectors in higher dimensions. For simplicity, we will write the single jump $P_1(l)$ as $p(l)$. To continue, we introduce a new function $\tilde{p}(k)$ (called a discrete Fourier transform)

$$\tilde{p}(k) = \sum_l \exp(ikl)\, p(l) \tag{1.3a}$$

and

$$\tilde{P}_n(k) = \sum_l \exp(ikl) P_n(l) \tag{1.3b}$$

Multiplying Eq. (1.2) by e^{ikl} and summing over all at l yields

$$\tilde{P}_{n+1}(k) = \tilde{p}(k)^{n+1} \tag{1.4}$$

Inverse Fourier transforming, we have

$$P_n(l) = \frac{1}{2\pi} \int_{-\pi}^{\pi} \tilde{p}(k)^n \exp(-ikl)\, dk \tag{1.5}$$

We are interested in the asymptotic behavior for large n and begin by expanding $\tilde{p}(k)$ in powers of k for small k,

$$\tilde{p}(k) = \sum_l \left[1 + ikl - \frac{1}{2}k^2 l^2 + \cdots \right] p(l)$$

$$= 1 + ik\langle l \rangle - \frac{1}{2}k^2 \langle l^2 \rangle + \cdots \tag{1.6}$$

where the moments of the jump distribution are defined as

$$\langle l^n \rangle = \sum_l l^n p(l) \tag{1.7}$$

For simplicity, we let jumps to the right and left have equal probabilities making $\langle l \rangle = 0$. Here we only study the case when $\langle l^2 \rangle$ is finite. When it is not, then we enter a topic of later discussion, the world of self-similar "fractal" behavior where there is no characteristic length scale. For $\langle l^2 \rangle$ finite, we can approximate $\tilde{p}(k)$ as a Gaussian in k-space

$$\tilde{p}(k) \approx \exp\left(-\frac{1}{2}\langle l^2 \rangle k^2\right) \tag{1.8}$$

$\tilde{P}_n(k)$ is achieved by raising both sides of Eq. (1.8) to the nth power. Inverse Fourier transforming, going to the continuum limit by letting the limits of integration in Eq. (1.5) go to plus and minus infinity with no harm as $P_n(k)$ has a sharp maximum around $k = 0$ for large n, we find the famous result of the Central Limit Theorem that $P_n(l)$ is a Gaussian,

$$P_n(l) = \frac{1}{\sqrt{2\pi \langle l^2 \rangle n}} \exp\left(-\frac{l^2}{2\langle l^2 \rangle n}\right) \tag{1.9}$$

that when summed over all l gives unity, as the walker must be somewhere with probability one. If in play, absorbing boundaries will allow for the walker to escape giving a probability less than one. In some cases, for specific random walks with a finite mean squared displacement, exact results can be obtained, but asymptotically one would still get the Gaussian.

Just for fun, let's consider a different approach, a rate equation with jumps at a constant rate λ, of length $+1$ or -1. Then we can consider the differential equation in time, rather than the difference equation in number of steps,

$$\frac{dP(l,t)}{dt} = -\lambda P(l,t) + \frac{\lambda}{2}[P(l+1,t) + P(l-1,t)] \tag{1.10}$$

which accounts for leaving site l at rate λ equally to the right and left, or coming to site l from the neighboring left-and right-hand sites each at rate λ. We can seemingly complicate matters by multiplying both sides of Eq. (1.10) by z^l and defining

$$G(t, z) = \sum_{l=-\infty}^{\infty} P(l, t) z^l \qquad (1.11)$$

which yields the equation for $G(t, z)$

$$\frac{dG(t, z)}{dt} = \lambda \left[-1 + \frac{1}{2} \left(z + \frac{1}{z} \right) \right] G(t, z) \qquad (1.12)$$

with the solution,

$$G(t, z) = \exp(-\lambda t) \sum_{l=-\infty}^{\infty} I_l(\lambda t) z^l \qquad (1.13)$$

where $I_l(\lambda t)$ is the Bessel function of imaginary argument and $P(l, t)$ will be the coefficient of z^l in that equation, i.e.,

$$P(l, t) = \exp(-\lambda t) I_l(\lambda t) \qquad (1.14)$$

Expanding $I_l(\lambda t)$ for large t and l values finds,

$$I_l(\lambda t) \sim \frac{e^{\lambda t}}{\sqrt{2\pi\lambda t}} e^{-\frac{l^2}{2\lambda t}} \qquad (1.15)$$

and thus,

$$P(l, t) \sim \frac{1}{\sqrt{2\pi\lambda t}} e^{-\frac{l^2}{2\lambda t}} \qquad (1.16)$$

and we get back asymptotically to the Gaussian. Here l is a dimensionless number and $\langle l^2 \rangle$ equals unity for nearest-neighbor jumps of length one.

Many different random walks have complicated solutions, but asymptotically assume the Gaussian form which is also the solution to the continuum diffusion equation,

$$\frac{\partial P(x, t)}{\partial t} = D \frac{\partial^2 P(x, t)}{\partial x^2}$$

whose solution is Eq. (1.16) with $D = \lambda/2$ having units of $[x^2/t]$ with x treated as dimensionless here. The diffusion equation approach can be expanded to several types of scenarios with particles that interact. This is the field of reaction diffusion equations that Turing initiated to describe pattern formation. Another example could be annihilation of diffusing particle (A) and antiparticle (B),

$$\frac{\partial P_A(x,t)}{\partial t} = D\frac{\partial^2 P_A(x,t)}{\partial x^2} - kP_A(x,t)P_B(x,t)$$

$$\frac{\partial P_B(x,t)}{\partial t} = D\frac{\partial^2 P_B(x,t)}{\partial x^2} - kP_A(x,t)P_B(x,t)$$

A's and B's near each other annihilate, leaving A rich and B rich regions which only find each other at their boundaries resulting in a slow decay of those in the interior of the A and B rich clusters. We won't go further into reaction–diffusion problems.

As another single-particle random walk example, consider random jumps in 2D of size a, but with equal probability in all directions. The former equations (1.3)–(1.5) still hold now even for a continuous 2D space r instead of for a lattice.

$$\tilde{p}(k) = \int_0^\infty d\boldsymbol{r} \int_0^{2\pi} e^{i|k|r\cos\theta}\delta(r-a)d\theta$$
$$= J_0(|k|a) \tag{1.17}$$

so, with k and r being 2D vectors and J being a Bessel function of zeroth order,

$$P_n(r) = \frac{1}{4\pi^2}\int e^{-i\boldsymbol{k}\cdot\boldsymbol{r}}|J_0(|k|a)|^n d\boldsymbol{k} \tag{1.18}$$

we can be confident that this non-obvious expression will for large n become

$$P_n(\boldsymbol{r}) = \frac{1}{2\pi\sigma^2 n}e^{-\frac{r^2}{2\sigma^2 n}} \tag{1.19}$$

the 2D diffusion equation solution with mean squared displacement $\sigma^2 = a^2$. If only interested in asymptotic properties when the mean squared jump distance is finite, the Central Limit theorem will prevail to converge to a Gaussian function no matter how complicated the probability looks, such as above containing a Bessel function.

The question arises is how good an estimate is the Gaussian after n steps. For a symmetric random walk only the k^2 term in the $\tilde{p}(k)$ expansion leads to the Gaussian. Differences come from higher order terms. For a symmetric random walk the odd moments are zero, so the Berry–Essen theorem looks at the absolute third moment to find the condition for the number of steps n needed to have the Gaussian be a good approximation around the central peak

$$n \sim \frac{\langle |l^3| \rangle^2}{\langle l^2 \rangle^3} \tag{1.20}$$

Here, and throughout the book, we ignore the mathematical rigor found in proper probability treatises.

1.2. Moments

Next, we look at moments. Staying in 1D for simplicity, we can calculate the moments of $P_n(l)$, for example the mean and mean squared position after n steps are

$$\langle l_n \rangle = \sum_l l P_n(l) = -i \frac{\partial}{\partial k} \sum_l e^{ikl} P_n(l) \tag{1.21a}$$

$$\langle l_n^2 \rangle = \sum_l l^2 P_n(l) = -\frac{\partial^2}{\partial k^2} \sum_l e^{ikl} P_n(l) \tag{1.21b}$$

when after the derivatives with respect to k are taken, then k is set equal to 0. The Nth moment would be found by taking N derivatives with respect to k, to bring down the factor i^N then setting $k = 0$ and multiplying by i^m with m chosen to make $i^{N+m} = 1$.

1.3.   Lattice Green's Functions

The Fourier transform is one kind of transform, here we introduce another where we sum over the number of steps,

$$G(l, z) = \sum_{n=0}^{\infty} P_n(l)z^n \qquad (1.22)$$

If we go back to Eq. (1.2) and multiply each side by z^{n+1} and summing over all n for a walker that starts at the origin, we obtain

$$G(l, z) - z \sum_{n=0}^{\infty} G(l', z)p(l - l') = \delta_{l,0} \qquad (1.23)$$

$G(l, z)$ is called a Green's function and it captures the behavior and properties of a single random walker (the source) starting at the origin. If $G(l, z)$ can be calculated then $P_n(l)$ is the coefficient of z^n. The Green's function will play a prominent role in calculating the probability when z is replaced by a transformed waiting time between jumps when time is a continuous variable. We first show here where the Green's function enters first passage times. We start by Fourier transforming both sides of Eq. (1.23) to obtain

$$G(k, z) = \frac{1}{1 - zp(k)} \qquad (1.24)$$

By the way, George Green developed his "Green's" functions to solve potential problems in electricity and magnetism. His Green's theorem transformed differential equations into integral equations whose kernel is a Green's function. Of course, he did not use the Dirac delta function, so familiar to us today, when applying the Green's function method.

1.4.   First Passage Times

We can also write an equation for $P_n(l)$ differently, but equivalent to Eq. (1.2) by invoking the concept of a first passage time $F_n(l)$ for reaching site l for the first time on the nth step,

$$P_n(l) = \sum_{m=0}^{n} F_{n-m}(l)P_m(0) + \delta_{n,0}\delta_{l,0} \qquad (1.25)$$

This equation has the random walker reaching site l for the first time on the $(n-m)$ step and then having zero displacement for the next m steps, so the walker takes the last m steps and is back at l on the nth step. The delta functions mean that the walker started at the origin. We next multiply both sides of Eq. (1.25) by z^n defining the first passage time generating function,

$$F(l, z) = \sum_{n=0}^{\infty} F_n(l)z^n \text{ and sum over all } n \text{ to obtain,}$$

$$G(l, z) = F(l, z)G(l = 0, z) + \delta_{l,0}$$

and solving for the generating function $F(l, z)$ to obtain,

$$F(l, z) = \frac{G(l, z) - \delta_{l,0}}{G(l = 0, z)} \qquad (1.26)$$

1.5.   Number of Different Sites Visited in n Steps

One of the first things for which we need first passage times, is to calculate how many different sites S_n on a lattice are visited in an n-step random walk. Starting at the origin and counting the origin as the first new site, we write,

$$S_n = 1 + \sum_{l} [F_1(l) + \cdots + F_n(l)] \qquad (1.27)$$

where $F_0(l) = \delta_{l,0}$. Forming the generating functions, ignoring the zeroth step as visiting the origin,

$$S(z) = \sum_{n=1}^{\infty} S_n z^n$$

$$= z \sum_l F_1(l) + z^2 \sum_l [F_1(l) + F_2(l)] + \cdots$$

$$+ z^n \sum_l [F_1(l) + \cdots + F_n(l)] + \cdots$$

$$= \frac{1}{1-z} \sum_l [z F_1(l) + \cdots + z^n F_n(l) + \cdots]$$

$$= \frac{1}{1-z} \sum_l F(l, z) \tag{1.28}$$

In Eq. (1.22), we note that $\sum_l G(l, z) = \sum_l \sum_{n=0}^{\infty} P_n(l) z^n = \frac{1}{1-z}$, letting us write Eq. (1.28) as

$$S(z) = \sum_{n=0}^{\infty} S_n z^n = \frac{1}{1-z} \left(\frac{\sum_l G(l, z) - 1}{G(l = 0, z)} \right)$$

$$= \frac{1}{(1-z)^2} \left(\frac{z}{G(l = 0, z)} \right) \tag{1.29}$$

The next step is to calculate $G(l = 0, z)$ then S_n will be the coefficient of z^n. We'll start with the 1D case

$$G(l = 0, z) = \frac{1}{2\pi} \int_{-\pi}^{\pi} \frac{d\theta}{1 - z p(\theta)} \tag{1.30}$$

for nearest-neighbor steps of equal probability to the right and left

$$G(l = 0, z) = \frac{1}{2\pi} \int_{-\pi}^{\pi} \frac{d\theta}{1 - z \cos(\theta)} \tag{1.31}$$

for $z < 1$ this is a known integral,

$$G(l = 0, z) = \frac{1}{\sqrt{(1 - z^2)}}$$

$$= \frac{1}{\sqrt{(1 - z)(1 + z)}} \tag{1.32}$$

How can we find the nth coefficient of z^n. Let's take a diversion with Fourier transforms. Consider,

$$f(t) = \int_0^\infty e^{-st} f^*(s) ds$$

with $f^*(s)$ varying as $s^{-1-\beta}$ as s $\to 0$ and $\beta > 1$ and rewriting the integrand around $s \to 0$, as

$$\int_0^\infty e^{-st} s^{-1-\beta} dt = \int_0^\infty t^\beta e^{-st} (st)^{-1-\beta} d(st)$$

$$= t^\beta \int_0^\infty e^{-x} (x)^{-1-\beta} dx$$

then for large t, $f(t)$ varies t^β when $f^*(s)$ varies as $s^{-1-\beta}$. We will come back to this when we treat random walks with a continuous time variable (the so-called continuous time random walk or (CTRW) but we explored it here, as there is a comparable technique called a Tauberian theorem that will allow us to find the coefficient of z^n in Eq. (1.29). The theorem goes as follows (ignoring factors of slowing varying functions such as logs):

When $S(z) = \sum_{n=0}^\infty S_n z^n = \frac{z}{(1-z)^2} \left(\frac{1}{G(l=0,z)} \right)$ *which goes as* $\frac{1}{(1-z)^\beta}$ *as* $z \to 1$, *then* S_n *goes as* $\frac{n^{\beta-1}}{\Gamma(\beta)}$ *for large n. In our 1D case* $S(z) = \frac{1}{(1-z)^{3/2}}$ *which, using the Tauberian theorem, yields for large n,*

$$S_n \propto \sqrt{n} \tag{1.33}$$

We did not prove the Tauberian theorem, but in our hand-waving fashion (used throughout) one can intuit that it is a variant of the dimensional analysis in our Fourier example. The $\sqrt{n}$ result is actually obvious. In the continuum limit, the random walk is governed by the diffusion equations (see Eq. (1.16) whose solution has a mean squared displacement $\langle r^2(t) \rangle \propto t$, so $r \propto \sqrt{t}$. In 1D for nearest-neighbor steps all sites between the walk end-points are visited and r corresponds to S_n and t corresponds to n. This argument won't work in higher dimensions as the trajectory won't be space filling. In 2D and 3D square and cubic lattices, one needs to calculate the Green's functions in the limit of $z \to 1$,

$$G(l = 0, z) = \frac{1}{(2\pi)^2} \iint_{-\pi}^{\pi} \frac{dk_x dk_y}{1 - z[\cos k_x + \cos k_y]} \sim -\frac{1}{\pi} \log(1 - z)$$

and

$$G(l = 0, z) = \frac{1}{(2\pi)^3} \iiint_{-\pi}^{\pi} \frac{dk_x dk_y dk_z}{1 - z[\cos k_x + \cos k_y + \cos k_z]}$$
$$\sim 1.51638 + O(\sqrt{1 - z})$$

The triple integral can be solved exactly for SC, BCC, and FCC lattices in terms of elliptic integrals and gamma functions. These are known as Watson integrals. The results lead to

$$S_n \propto \pi \frac{n}{\log n} (2D) \tag{1.34}$$

$$S_n \propto \frac{n}{G(0, 1)} (3D) \tag{1.35}$$

More generally, $S_n \sim n^{d_s/2}$ where d_s is called the spectral dimension. While d_s coincides with the lattice dimension for a regular lattice, it will be different for lattices with a fractal nature, such as a percolation lattice. This type of lattice has deadends, so once on a deadend the walker is cut off from the full lattice and fewer sites are visited. For percolation $d_s \sim 4/3$.

1.6.   Return to the Origin: k Walkers

These Green's functions come into play again when calculating the probability of a random walker starting at the origin and returning to the origin. From Eq. (1.26)

$$F(l = 0, z = 1) = \sum_{n=0}^{\infty} F_n(l) = \frac{G(l = 0, z = 1) - \delta_{l,0}}{G(l = 0, z = 1)}$$

$$= 1 - \frac{1}{G(0, 1)} \tag{1.36}$$

Since $G(0, z = 1)$ is infinite in 1D and 2D, the random walker will always return to its origin. In 3D on a simple cubic lattice, the walker returns with probability 0.34. These results are called Polya's theorem.

Let's focus on the 1D case. The walker always return to the origin, but after how many steps? The average number of steps $\langle N \rangle$ for a first return to the origin is

$$\langle N \rangle = \sum_{n=0}^{\infty} n f_n(0)$$

$$= \lim_{z=1} \frac{d}{dz} \sum_{n=0}^{\infty} f_n(0) z^n$$

$$= \frac{d}{dz} F(0, z = 1) = \lim_{z=1} \frac{1}{[G(0, z)]^2} \frac{d}{dz} G(0, z) \tag{1.37}$$

which diverges as $\frac{1}{\sqrt{1-z^2}}$ as z goes to one. The walker will always return to the origin, in 1D but this will take on the average an infinite number of steps.

What if there are there are k independent walkers, then the probability, in 1D, that none of the walkers has returned to the origin after n steps is,

$$Q_n^k(0) = \left(1 - \sum_{j=1}^{n} f_n(0)\right)^k \tag{1.38}$$

For one walker,

$$Q_{n-1}(0) - Q_n(0) = f_n(0) \tag{1.39}$$

that is, the probability that a walker has returned on the nth step is the difference between the $(n-1)^{\text{th}}$ step and the nth step probabilities. For the k independent walker case, the average number of steps for a walker to return to the origin is

$$\langle N \rangle = \sum_{n=1}^{\infty} n[Q_{n-1}^k(0) - Q_n^k(0)]$$

$$= \sum_{n=0}^{\infty} Q_n^k(0) \tag{1.40}$$

Converting Eq. (1.39) into a generating function equation by multiplying by z^n and summing from 1 to infinity gives

$$z \sum_{n=1}^{\infty} Q_{n-1}(0)z^{n-1} - \sum_{n=1}^{\infty} Q_n(0)z^n = \sum_{n=1}^{\infty} f(0)z^n = f(0, z) - 1$$

and solving for $Q(0, z) = \sum_{n=0}^{\infty} Q_n(0)z^n$ yields,

$$Q(0, z) = \frac{1 - f(0, z)}{1 - z}$$

then using Eq. (1.36)

$$Q(0, z) = \frac{1}{(1 - z)G(0, z)}$$

and in 1D

$$G(0, z) \sim \frac{1}{\sqrt{1-z}}$$

thus in 1D, $Q(0, z)$ diverges as $\frac{1}{\sqrt{1-z}}$, as z approaches 1, then the Tauberian theorem has $Q_n(0)$ behaving as $\frac{1}{\sqrt{n}}$ for large n. The sum in Eq. (1.40) goes as

$$\langle N \rangle \sim \sum_{n=1}^{\infty} \left(\frac{1}{\sqrt{n}} \right)^k \qquad (1.41)$$

So $\langle N \rangle$ diverges for $k = 1$ and 2, and converges when $k \geq 3$. So, in 1D, for $k \geq 3$ walkers, on the average one will return to the origin in a finite number of steps.

This exercise was to illustrate that one can ask an infinite number of questions about even a simple random walk. We will do one more many walker problem in Chapter 2 related to polymer physics.

1.7. First Passage Times Probability Distributions

Let's consider Eq. (1.25) in the continuum limit for a diffusion process in 1D,

$$p(x, t) = \int_0^t f(x, \tau) p(0, t - \tau) d\tau \qquad (1.42)$$

which we now Laplace transform $t \to s$ (for simplicity, we will not use an * to signify a Laplace transformed function or a $\sim$ to denote a Fourier transform x to k, so an s argument means there was a Laplace transform and a k argument means there was a Fourier transform to find

$$p(x, s) = f(x, s)p(0, s)$$

For the diffusion equation

$$p(x, t) = \frac{1}{\sqrt{4\pi Dt}} \exp\left\{ \frac{-x^2}{4Dt} \right\}$$

and Laplace transforming,

$$p(x, s) = \frac{1}{(4Ds)^{1/2}} \exp\left\{ -\left(\frac{sx^2}{D}\right)^{1/2} \right\}$$

so $p(0, s) = \frac{1}{(4Ds)^{\frac{1}{2}}}$ therefore,

$$f(x, s) = \exp\left\{ -\left(\frac{sx^2}{D}\right)^{\frac{1}{2}} \right\}$$

and transforming back into the time domain,

$$f(x, t) = \frac{|x|}{\sqrt{4\pi Dt^3}} \exp -\left(\frac{x^2}{4Dt}\right) \tag{1.43}$$

At first glance this might look like a Gaussian, but a Gaussian is peaked at the origin while $f(x, t)$ is peaked at $x^2 = 2Dt$, and note that for $x = 0$, $f(0, t) = 0$ (not peaked at the origin as being at the origin for the first time after a time t is zero since the walker started at the origin). This has a divergent first moment, so the mean first passage time to a chosen site is infinite. For example, if the chosen site is to the right of the origin and the walker drifts towards infinity on the left then the time to reach the site on the right will diverge. The probability density $f(x, t)$ as a function of t is called the Smirnoff distribution. It is an example of a Lévy distribution, a class of probabilities with infinite moments that we will encounter later in the discussion of Lévy flights and in anomalous diffusion.

For the lattice case the $t^{-3/2}$ will translate into an $n^{-3/2}$ behavior where n is the number of steps taken. Going back to Sec. 1.6 where in 1D

$$F(0, z) = 1 - \frac{1}{G(0, z)} = 1 - \sqrt{1 - z^2}$$

we take a z derivative, so we can apply the Tauberian theorem to

$$\frac{dF(0, z)}{dz} = \frac{z}{(1 - z^2)^{3/2}}$$

which diverges as $z \to 1$, as $(1 - z)^{-3/2}$ which converts to a $1/\sqrt{n}$ behavior.

As

$$F(r, z) = \sum_{n=0}^{\infty} f_n(r) z^n$$

$$\frac{dF(r, z)}{dz} = \frac{1}{z} \sum_{n=0}^{\infty} n f_n(r) z^n$$

so $n f_n \propto n^{-1/2}$ giving

$$f_n \propto n^{-3/2} \tag{1.44}$$

as we intuited from the Smirnoff distribution in the time-dependent continuous space case.

1.8. Number of Sites Visited in a Subset

We know how many different sites a random walker visits on a perfect lattice, but what if we ask for how many sites on a line, a distance m from the origin, are visited after n steps. Let's choose a set T and ask how many sites on T are visited in n steps.

$$S(z) = \frac{z \sum_{l \in T} G(l, z)}{(1 - z)G(0, z)} \tag{1.45}$$

Previously T was the whole lattice and the numerator would be $z/(1 - z)$. For a 2D lattice, let's choose the line of height m as the set T and ask for the mean number of different sites visited on the line after many steps n. T is the set (j, m), with $j = 0$ and m, zero and all positive and negative integers. The quantity of interest is,

$$\sum_{l \in T} G(l, z) = \frac{1}{2\pi} \int_{-\pi}^{\pi} \frac{\exp(-imk_x)}{1 - zp(k_x)} dk_x \tag{1.46}$$

We did not show the k_y integral over the terms with the j's as it is unity since the term $\sum_{j=-\infty}^{\infty} \exp(-ijk_y)dk_y = \sum_{j=-\infty}^{\infty} \delta(\frac{k_y}{2\pi} - j)$ and the integral over k_y goes from $-\pi$ to π, so only the $j = 0$ term contributes. For a symmetric random walk, in the small k limit

$$p(k) \sim 1 - \frac{1}{2}\sigma^2 k^2$$

where σ^2 is the single step variance. For $z \to 1$, and symmetric jumps we can rewrite Eq. (1.46) as

$$\frac{1}{2\pi} \int_{-\pi}^{\pi} \frac{\cos(mk)}{1 - z + \frac{1}{2}\sigma^2 k^2} dk$$

$$= \frac{1}{\sigma[2(1 - z)]^{1/2}} \exp\left\{-\frac{m}{\sigma}[2(1 - z)]^{1/2}\right\}$$

Using the 2D $G(0, z)$, we find

$$S(z) \sim \frac{1}{(1 - z)^{3/2} \ln(1/(1 - z))} \tag{1.47}$$

where, as $z \to 1$, we can neglect the exponential term and use the Tauberian theorem to find for different sites visited on a line in 2D

$$S(n) \propto \frac{\sqrt{n}}{\log n} \tag{1.48}$$

as compared to $n/\log n$ for sites visited on the whole lattice. Points visited on a line in 3D would only grow as $\log n$. This is just one more example of there is no end to questions that can be asked about random walks.

I did this work on subsets with George Weiss at NIH in 1982. George was an early grad student of Elliott Montroll and now was a senior researcher at NIH. Early in his career he solved a nonlinear centrifuge equation that allowed the relevant information to be extracted without needing to wait for asymptotic behavior to be reached. This fit the interests of NIH and George spent most of his career there, until his retirement in 2010. Visiting George one day, he suggested this problem which we worked out in his office, wrote it up, and had it published. George was an applied mathematician extraordinaire and a prolific writer. His book *Aspects and Applications of the Random Walk* is both modern and a classic. In those days, I would drive on to the NIH campus and park next to building 12A where George worked and walk upstairs one flight to his office. Times change. These days NIH is surrounded by an iron fence with heavy security at entrances, including opening the trunk of your car and mirrors on sticks to check underneath. A visitor badge is required from security.

CHAPTER 2

De Gennes' Reptation

2.1. Untying a Many Chain Knot Finding a 10/3 Exponent

In a polymer melt, all the polymer chains have the configuration of a random walk in 3D. We use the mass M of a chain as proportional to its length when stretched out as a line. In a melt, the chain's end-to-end distance is $\sqrt{M}$. The model is that a chain end moves one unit when a kink or defect on the chain performs a random walk and reaches a chain end. For a random walk between two ends separated by a distance M, the time to reach an end varies as M^2. For the chain, whose M links are in a random walk configuration, it will take M chain end movements for the chain to move its end-to-end distance of $\sqrt{M}$ In the reptation model, one freezes all the chains, but one, and calculates the time for that one chain to move its own length of $\sqrt{M}$. Each end movements takes a time M^2, so the diffusion constant varies as $D \sim 1/M^2$. M end movements are needed for the chain to move its length of $\sqrt{M}$ and each end unit movement takes a time varying as M^2, thus there is a time scale $T_R \propto M M^2 = M^3$, called by DeGennes to be the reptation time. However, experimentally, one finds

$$T_R \propto M^{10/3} \tag{2.1}$$

Let's ask a slightly different question of the DeGennes model. The chain is contained in a sphere of radius $R = \sqrt{M}$ and volume $M^{3/2}$. A knot of $\sqrt{M}$ entangled chains, each of mass M, fills this volume. The question now is how long will it take $\sqrt{M}$ chains to leave this sphere. It will certainly be longer than the M^3 time for a single chain to leave the sphere.

The following is from works with Harvey Scher, John Bendler, George Weiss and myself. Let's start that calculation by writing the probability, in 3D, of a single random walker being at site r, at time t, when starting at site r_0 at time $t = 0$.

$$p(r, t|r_0, 0) = 4\pi \left(\frac{3}{4\pi Dt}\right)^{3/2} \exp\left(-\frac{3(r - r_0)^2}{4Dt}\right) \qquad (2.2)$$

where Dt has units of r^2. Integrate $p(r, t|r_0, 0)$ over all r and r_0 from zero to $R = \sqrt{M}$ and let $P(t)$ be the probability that if a walker starts inside a sphere of radius R, then it is still inside that sphere at time t. For $Dt \gg R^2$, the double integral over r and r_0 gives

$$P(t) \sim \sqrt{2\pi}(R^2/Dt)^{3/2} \qquad (2.3)$$

which we write as $P(t) \sim (T/t)^{3/2}$ with $T \propto M^3$ as $R^2 \propto M$ and $D \propto M^{-2}$. The probability $S(t)$, for N non-interacting walkers starting in the sphere of radius R, that at least one of them is in the sphere at time t, is

$$S(t) = [1 - (1 - P(t))^N] \qquad (2.4)$$

and since $dP(t)/dt$ is negative as $P \sim t^{-3/2}$, then $-dS(t)/dt$ is the probability for the sphere to be empty. The mean time $\langle T_N \rangle$ for the sphere to be empty (not necessarily for the first time of the N random walkers, but that's not a bad assumption in

3D) is

$$\langle T_N \rangle = - \int_0^\infty t \frac{dS(t)}{dt} dt = \int_0^\infty S(t) dt \qquad (2.5)$$

where we have used integration by parts. For large N,

$$\langle T_N \rangle \sim \int_0^\infty (1 - \exp(-NP(t))) dt$$

$$\sim \int_0^\infty \left(1 - \exp\left(-N\left(\frac{T}{t}\right)^{\frac{3}{2}}\right)\right) dt$$

$$\sim T N^{2/3} \qquad (2.6)$$

For $N = \sqrt{M}$ entangled chains, and $T = M^3$ the reputation time for $N = 1$ single chain to move its own length, then the time to disentangle a knot in a polymer melt,

$$T_{N=\sqrt{M}} \sim M^3 M^{1/3} = M^{10/3} \qquad (2.7)$$

a longer time than for a single chain as some chains may move in the same direction and stay entangle at time M^3. This illustrates one example of fractional exponents arising from Brownian type motion, by asking a many walker question of a standard random process. Fractional exponents can appear when the random process has an explicit fractal space or time quality.

I sent the above calculation to Pierre Gilles DeGennes and on April 30, 1986 he wrote to me, "On the reptation idea, the notion that we need $\sim \sqrt{M}$ chains to move out to relax one regime is very novel to me but extremely stimulating" signed Pierre G de G.

I met DeGennes for the first time at the November 1986 Materials Research Society meeting, in Boston, where he gave a talk, the last one of the session. Afterwards he was surrounded by several people wanting to speak with him. I waited patiently until only I was left and I said, meeting you is reaction-limited, not diffusion-limited, as it was easy enough to reach him, but there was a long delay in getting to speak with him. He laughed, saw my name tag and said, Shlesinger let's go to lunch I want to talk with you. He found a French restaurant. DeGennes could make you feel that you were the famous physicist and he was the student. At one meeting in Cargese, Corsica I was sitting in the back with friends making sarcastic remarks about the speakers. When DeGennes finished his talk, he walked back and said to me, can I join your gang of loiterers. Getting back to our lunch, he said his theory predicts M^3, the theory that predicts $M^{10/3}$ is mine. I said, your reptation theory works perfectly, you just need to ask a different question of the model. Anyway, despite the fact that the experimental result has been obtained, the polymer community has ignored this work and seems to still be accepting of the single chain M^3 result. Perhaps, someone will revive this point and look further into polymer melt relaxation dynamics, not just the exponent, but also the prefactor.

2.2.   Jekyll and Hyde Referee Report

Below is the referee report for our reptation paper "Polymer Melt Dynamics Model with a Relaxation Time Exponent of 10/3", George Weiss, John T. Bendler, and Michael F. Shlesinger, *Macromolecules* vol. 21, 521–523 (1988). The report is a humorous take on Dr. Jekyll and Mr. Hyde, with the good doctor J having positive comments and his awful colleague Mr. H having negative comments. I believe the referee is Pierre-Gilles, but I never asked.

25 July 1987

Re: Macromolecules #87-476
    G. H. Weiss, J. T. Bendler, and M. F. Shlesinger
    "A polymer molecule with a reptation time exponent of 10/3"
    Referee I

Mon cher collègue:

    You put me very much in your debt with this opportunity to review a manu-
script for you. Yet I must confess with embarrassment that my schedule does not
afford me an opportunity to read the literature. I have had to pass this honor
on to two of my associates. Their remarks were naturally not returned to me in
English and I have had to translate them as best I could. To help you in judging
their remarks I have thought it best to comment frankly on their qualifications.
    Monsier H. is a scoundrel of uncertain lineage. Mon Dieu! He insults even
me. I keep him with me only because I don't know how to get rid of him. (By the
way, are you taking graduate students?)
    Doctor J. is, how do you say it, another kettle of bouillabaisse. A mature
fellow, regretably without ambition, certainly not an athlete. I would trust him
with my mistress. You should follow his recommendation.

M. H.

Professor:
    These people come across an exponent, and pray that the physics will be
found. What an abomination! [Quelle abomination!] The Professor himself does not
escape a measure of responsibility for such things.
    The authors claim in the abstract to find a reptation time varying as M^{3+x}
with $x=1/3$, but this is nonsense, at best an attempt to masquerade their anti-
theory in the seductive guise of the reptation model. It is well known that the
reptation time varies as M^3. If they wish to claim a different power they should
associate it with a different concept.
    Neither these authors nor those they follow (H. Scher and M. F. Shlesinger,
J. Chem. Phys. **84**, 5922 (1986)), make a creditable attempt to relate their many
particle relaxation time to stress relaxation. It is perfectly obvious that
stresses arise and are maintained in the backbone structure of each chain, as is
commonly supposed even in the theory of "phantom" chain rubber elasticity, and
also in all detailed models of polymer melts (Doi and Edwards, Bird and
Curtiss,...). Therefore if only one chain loses its "network junctions", that
chain will relax its stresses. Why, how, do all these other chains within a
sphere of radius $M^{1/2}$ enter the picture? Why, how, do they diffuse independently, if they all participate in each others' junctions? What does it all have to
do with stress relaxation?

The question is not whether the editor would want to publish this, but why would the authors?

As to the details:

Eq. (7): This is not obvious. A gaussian probability function of the vector $r-r_o$, which gives the conditional probability density for passage from r_o (on the z axis, say) to r, will have after integration over the spherical angle coordinates of r two exponentials, one with $(r-r_o)^2$, the other with $(r+r_o)^2$.

Eq. (11): This has a digitalgraphic [?] error $(3/2 \to 2/3)$

Dr. J.
Professor:

The authors wish to bring to our attention a quite interesting phenomenon, a many body exponent which is a rational number, unlike so many. I can certainly see no reason why the requirement that all chains diffuse out of a sphere of size $M^{1/2}$ is irrelevant to melt dynamics. And if the authors have not yet spelled out in excruciating detail what the relevance is, maybe they will, or maybe someone else will, once it becomes known that some people consider this to be an important goal.

As to Monsier H., what an abomination! [Quelle merde!]. If he's so smart why doesn't he write a paper himself instead of insulting the sincere efforts of his moral and intellectual superiors!

This paper should of course be published. Perhaps there is no need to grant the authors the haste which they so understandably desire. But publish in the fullness of time.

CHAPTER 3

IBM and My First Random Walk Experience

The World's Fair opened in Flushing Meadows in Queens, New York in 1964. I visited several times. It was a wondrous experience. Major corporations had exceptional exhibits, with rides, shows, and an emphasis on predicting the future of technology. IBM had an exhibit with a computer that played Tic-Tac-Toe, many displays and features, but what I remember the best was a poster that explained a random walk. It used the analogy of a drunk starting at a lamp post and wandering around, the so-called drunkard's walk. There was a mention of Einstein's work. I don't remember the details, but it probably discussed Brownian motion and Einstein's method for illustrating the existence of atoms. Benoit Mandelbrot was at IBM and now I wonder if that random walk poster was due to him. In all my years interacting with Benoit, I never thought to ask. Anyway, the discussion of the mathematics of probability and random walks permanently drew my interest to that field. By the way, the food at the World's Fair featured then the famous Belgian Waffle that was topped with fresh whipped cream and strawberries. It was a sensation! At one of the restaurants there was a soda dispenser where one could select from a variety of choices and fill up a paper cup and get refills by yourself. This was a completely new experience. As I said, the Fair was predicting the future. At that time, in a restaurant, when one ordered a soda, it was drawn from behind a counter by a "soda jerk" and served in a glass. The joke, in those days, was to order a soda in a clean glass.

My love of math and my reading the IBM poster about a random walk, led me to take a probability course my senior year in high school which was pure fun solving problems about permutations and combinations and other probability puzzles. I also took a probability course in college, but instead of problem solving it focused more on questions of structure and questions about Weiner measures for mathematically analyzing the trajectory of a Brownian particle. The trajectory is continuous, but nowhere differentiable and the distance between any nearby points on the trajectory is infinite because the trajectory fills two dimensions so a one-dimensional measure is unbounded. Lewis Fry Richardson wrote about the boundary between Spain and Portugal changing with the minimum length scale used. In this way, Portugal reported a longer boundary. Later Mandelbrot wrote about the coastline of the UK being infinite in the limit of a decreasing ruler size. I didn't know in college that the Brownian trail was an example of a curve with fractal properties. Mandelbrot brought the knowledge and excitement of fractals to scientists and the public in his groundbreaking book, The Fractal Geometry of Nature. His works had a great influence on me. The Brownian curve was said not to be of bounded variation. In my real analysis course all the theorems had the stipulation that curves were of bounded variation, omitting all discussion of fractal curves. We did study the Cantor set where each point is a limit point but the set is nowhere connected. But this was not used as an introduction to fractals, but used as a cautionary tale that math required rigorous proofs because intuition can fail you, as you might assume if every point in a set is a limit point the set must be connected somewhere. That's why mathematicians want proofs, not intuition, and why you are judged not by how many problems you solve, but by how theorems you prove. Theoretical physics weighs in on the side of how many problems you solve. I understand abstraction. You might start learning that three apples and two apples equals five apples. Later you say $3 + 2 = 5$ without the need for a specific item. But for me, I liked solving problems more than proving theorems, so I switched from math to physics for graduate school.

My love of math did not start with the IBM exhibit, but I believe was hardwired from birth. My father graduated from NYU Law School, in 1934, at the age of 21. He was brilliant and quick and knew my high school geometry, trigonometry and algebra, but his math stopped there. He never made it to calculus. He went into law and his cousin around the same age became a physician. The family wanted to cover the professions that could care for their well-being. My father's parents emigrated from Russia in early 1906 as part of a larger emigration from turbulent times. My grandfather and his brothers opened a light fixture store to supply the many skyscrapers that were being built in New York. In 1929, my grandfather opened Hotel Shlesinger in Loch Sheldrake in the Catskill mountains in upstate New York. In 1942, my father enlisted in the Army, trained in Panama, shipped out to Australia from California, and in 1945 ended up in the liberation of Manila, where afterward he was assigned to a ship for the invasion of Japan. Thankfully that never happened, but he came home with malaria with a long recovery period. My twin brothers were born in 1946 and my father lived to 97, retiring at 60 and, in my memory, was never sick. Perhaps, the struggle with malaria boosted his immune system? His parents needed help managing the hotel, so my father and his younger brother joined their parents to manage the hotel. That was only from Memorial Day to Labor Day, but sufficient profit was made to retire to Brooklyn for the rest of the year. Before WWII they spent the winters in Miami Beach.

I was born on August 8, 1948 during a heat wave. My mother left Hotel Shlesinger on August 7 to give birth to me at Brooklyn Jewish Hospital where she worked as a nurse during the months when the hotel was closed. The rest of the family joined us at the end of the 1948 season, where we lived in Sheepshead Bay in rental housing for veterans. Today I believe it is close to the site of a community college, nearby to Coney Island. You might know that Coney Island is not an island, but apparently, it was back in the 1600s before its narrow inlet silted over. And the island was a favorite for rabbit hunting and Coney was what sounded like "rabbit" in Italian. My memories are at 3 years old walking with my brothers to their kindergarten and playing with toys until my mother left with me. On weekends the outdoors was flooded with children, playing games like shooting marbles out of circle. It was not unusual to see people with polio in wheelchairs or crutches. There were no vaccines or antibiotics and kids ran through the gamut of measles, mumps, and chicken pox. I don't remember getting ear or throat infections, however when my brothers came down with the measles, my mother made sure I caught it from them. No quarantines in those days. My memory is that it was fun to get red spots all over my body. Today, of course, a measles epidemic would be a cause for alarm. We got mercury fillings and our dentist gave us mercury to place in our palms to watch it roll around. What a difference today. There was a report of a mercury thermometer dropped in a high school science classroom forcing the evacuation of the room.

The Hotel Shlesinger was a wonderland for me, but probably hard work for my parents. There was the main house with guest rooms on upper floors, and on the ground floor, reception, offices, a game room, adult and children's dining halls, the kitchen, a bakery where I knew the timing when the cookies came out of the oven. Outside was a playground, a summer camp, cottages, bungalows, a casino with a house band, and up on a plateau was the swimming pool. On one side was a cow pasture (today it's a golf course) and across the street was a strong smelling chicken farm. On the pathway to the main house card tables were set up and the men played chess. By almost osmosis I learned the game of chess just by watching. My goal was one day to beat my father in chess. One day I walked into my father's office and he had a large desk calendar. I do not remember how old I was, but pre-school, and I remember for the first time realizing that the days were arranged into weeks. Before that, the days just rolled into each other with no pattern. I was happy to see the numbering matrix. We sold the hotel in 1954 getting ahead of the decline of the Catskill Mountains as the highways, by 1960, opened the country up to long drives, motels, plus air travel became feasible. New Yorkers had too many other opportunities to remain tied to the Catskills. My father and his brother in 1955 opened up a very successful toy and children's furniture store, Toy World, in the first outdoor mall, the Cross County Shopping Center, located in Yonkers, a suburb of New York City. It was not far from the 241st subway stop in the Bronx which was a direct line to 161st the Yankee Stadium stop.

My mother's family ran a grocery store in Hurleyville, a hamlet in the Catskills mountains. They supplied the hotels, such as Shlesinger's in nearby Loch Sheldrake. My mother, as a child, would go along with the deliveries to the hotel owned by my

father's family. New York gangsters, known as Murder Incorpo-
rated, also frequented the Catskill mountains in the summer and
the grocery store of my mother's family. When they came, I was
told that my grandmother would send my mother and her sister
upstairs to their apartment. My mother said the gangsters were
always polite and paid their bill, but supposedly buried bodies
in upstate New York. When I went to grad school in Rochester,
I learned that the rest of New York considered the Catskills as
suburban NYC and not upstate NY as it was called in NYC. My
mother's father was said to be good with numbers and passed
away from tuberculosis when she was 3. Her mother passed away
13 years later. She and her sister went to live with their aunt
Rose in Brooklyn. Rose's grandson became a physicist working
at UC Berkeley and then MIT's Lincoln Labs. In 1965 I was
applying to colleges, he told me that a good choice would be
SUNY Stony Brook on Long Island, a place I never heard of. He
said a group of Princeton physicists were joining and it was an
up-and-coming place. I applied and was accepted. I scored well
on the NY State Regents Exam and received a Regents scholar-
ship. The amount was based on parent's income and I got the
lowest amount of $350 a year, but tuition then was only $400 a
year. My cousin (my mother's sister's son) also took the Stony
Brook advice and came to major in electrical engineering. We
were roommates for four years and he became an electrical engineering professor at RPI designing turbines by employing the
first principle Maxwell equations to handle eddy currents. His
family had a bungalow colony in Hurleyville, New York (down
the road from Loch Sheldrake) that they operated for many
years. A fair number of talented people have come from these
small hamlets and towns with the proverbial one room school
houses.

By the way, back in the mid-1960s college bound students
took the SAT an aptitude test with two sections, math and

verbal. We were told that it was an intelligence test, so just show up and take it. So, one Saturday morning, you came in, took the test and went home. Today there is an industry of study guides and tutors to prepare students for the SAT, taking it as many times as needed to reach a high enough score. This is now a high pressure, stressful experience, as with many other things making life more complex. Actually, studying could have helped me. On my test, there was a math problem that gave an $f(x)$ and a $g(x)$. The problem asked to find $f(g(x))$. Although trivial for me today, I did not understand what a function of a function meant back then. Much later, in 2003 I wrote a paper with colleagues on the solution of $f(f(x)) = x^2 - 2$. Although this equation looks simple enough it is challenging to solve and involves a cosine and an arccosine. Supposedly the problem was posed by Brian Josephson of superconducting junction fame.

Apparently, Alfred Binet created what we call the IQ test as a means for a teacher to discover areas in which a student would need extra focus. His friend Poincare examined the math questions and brought up points that could be open to interpretation. Consider the sequence, 1, 2, 3, 4. What is the next term? While 5 seems an obvious answer, the number -19 could also work if the sequence is generated from the function

$$f(n) = n - (n - 1)(n - 2)(n - 3)(n - 4).$$

Deep thinkers could find themselves in trouble on a SAT exam. Apparently, Poincare pointed out to Binet examples of alternative non-obvious solutions to math questions on the test.

The verbal part of the SAT could also try an inventive mind. A type of question on the verbal part is, one reads a paragraph and then picks from multiple choice answers, something like (a) the industrial revolution, (b) women's rights in England, (c) international commerce, (d) personal courage. No chance to defend your answer, you just needed to pick the right one. I found the verbal part unsatisfying, but the test was over and quickly forgotten. Not like today where typically a student takes the test several times and submits the best result.

Nelson Rockefeller was the governor of New York and the driving force behind expansion of the State University of New York system. He was on campus for the groundbreaking of an Engineering building at Stony Brook. There were speeches and then Rockefeller climbed onto a bulldozer to move the first bit of earth. It was put in gear and suddenly jerked backwards, startling the people behind. The driver next to him quickly corrected the mistake. Much later, I later read that Rockefeller was dyslexic.

Back in my high school, calculus was mentioned in hushed tones, as if it was some incredibly difficult subject, despite Newton's Principia of 1689 being nearly 300 years old. I was overjoyed, as a Freshman, using the Protter and Morrey calculus book, to discover that it was so easy to learn. I was disappointed to find the higher-level math courses were about mathematical structures, proving theorems, and not about problem solving. As a Math major, I moved my attention to physics, taking six physics courses in my senior year, including an electronics lab, to finish with a dual degree in math and physics. The Physics Department was a community that included the undergrads, while the Math Department seemed to be a collection of loners. I don't remember there being a math club and I don't remember there being or going to a math colloquium. I did go to most of the physics colloquia.

I was also disappointed in the two economics courses that I took. Neither course used much math other than simple algebra.

Lines of supply and demand were straight, so no calculus was enjoined. One economics professor said he never learned calculus. I also took a course on computer operating systems, where to store, move, and retrieve zeros and ones, that I thought was dull. Little did I comprehend the operating system, like DOS, was where the money was.

My cousin and myself were studying for second semester calculus final. One of the problems he suggested trying was cutting a slice from a circle to create a cone with maximum volume. We puzzled over it and solved it. That turned out to be the big point problem on the final exam, that I finished in half the allotted time. The next year we were both placed in an honors junior level math course. By the way the Chairman of the Math Department was James Simon, who later founded the billion-dollar hedge fund, the Renaissance Fund and donated millions to Stony Brook's Simons Center for Geometry and Physics.

The Physics Department had an NSF grant with summer support for undergraduate students. I guess no one else wanted to stay on campus for the summer, but I jumped at the chance. I asked to work with a theoretical physics professor, but he declined. He said an undergrad sophomore wouldn't be useful. But having worked through most of the Feynman Lectures Volumes I and II, I felt I was ready for anything. I ended up with the Nuclear Physics Van der Graaf Lab which was under construction that summer. I learned to set the energy and optimize the beam and worked shifts. However, my main work was programming, the PDP-9 and the IBM 360 mainframe. The PDP-9 came with an instruction book that I was the first to read and ended up writing many programs for the lab. The computer started with a paper tape and toggle switches. But it had a monitor and a light pen that one could shine on one of nine sections on the screen to pull up subroutines to analyze data different ways. As the underlying math was binary, I wrote a program for the light pen to plot data in base 8. I was asked to change it to base 10.

One of my physics colleagues approached, Max Dresden, a lively physics professor to ask for some additional reading. Dresden did not teach any undergraduate courses, but was known to us through his enthusiasm for physics, including meeting with the physics club. He suggested that my friend read Michel Loeve's book on probability. I had enough on my plate the last semester to not ask for additional work. I was taking three physics courses, a math course, and an economics (or maybe it was a computer course) plus I was working at the Nuclear Structures Lab running experiments and programming. I took the physics GRE which didn't seem hard, and I went with my friend to ask Dresden for advice about graduate school. He drew an imaginary map of the United States in the air and proceeded to discuss possibilities. The one that stuck in my mind was a mention that an Elliott Montroll, at the University of Rochester, had won a large NSF grant to create a center for applied math and his field was statistical mechanics. He didn't say statistical physics, but in his Dutch accent placed the accent strongly on the "me" in mechanics. Dresden's suggestion turned out to be heaven sent and shaped the rest of my life. I went to Rochester and my friend went to Cornell. By the way, my cousin's niece just did her PhD at Cornell, so something is running through the genome.

CHAPTER 4

Stony Brook and the Laughing Dirac

My high school physics teacher said that physicists pay more attention to the equations than to the words describing them. In high school, we had Newton's laws and Kepler's laws, but not the equations for them, just the words. In college that all changed as the results were all shown to be derived from an inverse square force law. I started at SUNY Stony Brook in 1966 as a math major, but I also signed up for freshman physics. My teacher was Arnold Strassenberg, an exceptionally good lecturer, and we used the Berkeley series book on mechanics and he suggested the Feynman Lectures for supplemental reading. I fell in love with the Feynman Lectures. I never learned so much so rapidly and physics has never since been that easy for me. I ended up with a double math and physics major, took the physics GRE and only applied to physics graduate schools. Strassenberg told the story that on his PhD thesis defense, Feynman asked "what is the velocity of the electron in the first Bohr orbital?" When the answer was not forthcoming Feynman was upset. It, of course, was the speed of light times the fine structure constant, and it was the first place the fine structure constant appeared in physics. Despite the fact that one thinks of Bohr orbitals in terms of energies, Feynman found it hard to accept this constant's first appearance was not appreciated, as its small value allows for perturbation calculations in quantum electrodynamics. Its value is close to 1/137 and the Stony Brook physics lecture hall room is number 137.

Paul Dirac, one of the founders of quantum mechanics, was facing eventual mandatory retirement from Cambridge and came to spend a semester at Stony Brook, I assume to search for a position in the US. He eventually settled at the University of Florida. At the Nuclear Structure Lab where I worked, I overheard some graduate students speaking about his teaching method. He taught a graduate course in quantum mechanics by reading directly from his book. When asked why, he replied that he spent a great deal of effort choosing the right words in writing the book and didn't think he could improve upon his writing, hence he read the book to the class.

I joined the physics club and the six Feynman Baker Lectures from Cornell were shown over six nights on film reels. Dirac came to all six films and sat with the undergrads. Feynman had written about his own encounters with the famously taciturn Dirac about how hard it was to get a word out of him and certainly not a smile. Dirac was known for demanding beauty in the fundamental equations of physics. Beauty is not a requirement in the laws of physics, but Dirac felt strongly that the basic laws should possess a certain simplicity and elegance. In one of the Feynman films, he discusses in his thick New York accent, symmetry and beauty and mentions that there is this guy named Dirac who thinks this beauty thing is essential to good physics. I was sitting behind Dirac, and at this point I heard him release a good laugh. Finally, Feynman did get a reaction from Dirac. I have never heard any other story about Dirac laughing.

At the end of the semester, Dirac gave the physics colloquium. There was standing room only as people were expecting a folksy tale of the birth of quantum mechanics. Instead, it was a very dry talk directed towards the blackboard about the infinities in quantum field theory. The room slowly emptied out as Dirac bemoaned the hazards of field theory with particles popping out anywhere and being annihilated elsewhere, rather than a particle equation of motion. In the front row was CN Yang the Nobel laureate whose Yang-Mills fields were the foundation of quantum field theory. Yang started to defend field theory, when Dirac walked over handed him the chalk and sat down. Yang, clearly embarrassed, made a quick kind remark, and the colloquium abruptly ended, with no smiles or laughs.

Another interesting seminar was by James Watson, of DNA helix fame, coming from the nearby Cold Stream Harbor Lab. He had strong advice for grad students, to find their own thesis topic, as otherwise they will follow in their advisor's footsteps and the advisor would get the credit. He moved from the study of phages, to his own topic of DNA structure. I didn't follow his advice.

Finally, graduation came in the spring of 1970. The keynote speaker was Joseph Heller, the author of Catch 22. He started by saying that life was going to be filled with riches, travels, fame · · · "but enough about me" your life is going to hard. This was in the middle of the Vietnam war with graduates becoming eligible for the draft. It may have turned out for many us that the joke was on Heller, that indeed we may have had the better life. Certainly, there are enough worldwide water holes for physics conferences to surpass Heller's travels.

CHAPTER 5

Another Laughing Story: How to Solve the Quadratic Equation at the White House

I heard this story secondhand, but had it confirmed. John Deutch, a chemical physicist from MIT, became the Director of the CIA. At a Clinton Cabinet meeting he was sitting next to Larry Summers, the Harvard economist who was the Secretary of the Treasury. Looking around at the President and the cabinet members, supposedly Summers passed a note to Deutch which read, how many people here do you think could solve the quadratic equation. Deutch wrote a reply and passed the note back. It read "one". With Deutch from MIT and Summers from Harvard this was the equivalent of saying if a professor moved from MIT to Harvard the IQ of both places would rise.

By the way, I was never taught the geometric way to solve the quadratic equation, but this derivation seems easier than trying to memorize an answer. Never mind that b and c need to be positive, the answer works for all b and c.

$$x^2 + bx = c$$

Start by drawing an inner square of side x, the inner white box has area x^2, draw the four boxes of sides x and $b/4$. The four together have area of bx. Now complete the square (that's where the phrase comes from) so the large square has area $(x+b/2)^2 = x^2 + bx + b^2/4$ where on the right-hand side we have added up the individual pieces in the picture. Note from the equation $x^2 + bx = c$ so we can write $(x + b/2)^2 = c + b^2/4$ and solve for x as

$$x = -b/2 \pm \sqrt{c + b^2/4}$$

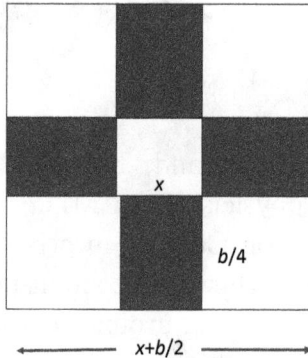

This geometric type approach also can be used to solve cubic equations by completing the cube.

CHAPTER 6

Pitfalls and Paradoxes
in the History
of Probability

6.1. La Jolla Institute Interlude

For graduate school, I was accepted to UCSD in La Jolla, a
sunny paradise, but I elected to go to the snow belt location
of the University of Rochester. Upon completing the PhD the
opportunity opened to work in La Jolla, and it became my home
away from home. I was even the UCSD Regents Lecturer in 1992,
nominated by Roger Dashen and Harry Suhl, two great theorists.
This connection to La Jolla was initially due to my thesis advisor, Elliott Montroll. He along with four other professors opening
a small non-profit company, in the early 1970s, called Physical
Dynamics. Apparently, an engineering professor decided to leave
the academic life and this was the start of the idea to open a company. The business plan was for classified work in a secure facility on projects with classifications at odds with open research
on a campus. An added benefit was that professors could hire
their best grad students, upon graduation, as research scientists,
instead of their moving to a university. Also grants could come

through Physical Dynamics instead of the university with the overhead spent on company expenses versus university administration. On my initial visit, there were dinners paid by a business credit card. I guess that's how non-profits work.

Montroll had been the Director of Physical Research at the Office of Naval Research in the 1950s, and later the Vice President of the Institute for Defenses Analyses in the 1960s and in the 1970s was managing the contract to the Center for Naval Analyses through the University of Rochester. He was also a member of the DOD advisory group then managed by DARPA, the JASON's, so he had many DOD connections that helped bring in grants from the DOD. Several such companies, like Physical Dynamics, began around this time and some became billion dollar successes, such as SAIC, but at the time it was not clear which would flourish. Physical Dynamics's business plan broadened out to an offshoot the La Jolla Institute (called by all LJI). My first visit was in 1975, following a soliton conference at University of Arizona in Tucson in February. At the Tucson meeting, it was warm and sunny during the day and I did not realize that the heat was turned off in my hotel room. When I woke up in the morning I had a sore throat and the room was freezing. Yes, the desert is cold at night. Afterwards, arriving in La Jolla was a welcome contrast.

LJI was located in McKellar Plaza on Prospect Street in La Jolla overlooking the "cove" (now populated by seals). Through LJI's large picture windows I watched a school of migrating whales swimming north. It was an idyllic scene. Traffic in La Jolla was so sparse one could cross the street, almost without looking. Meeting with Montroll outside of LJI, he looked up at some construction and said, they are ruining the place, and indeed La Jolla today is a place where one crosses the street carefully. LJI was something akin to an early version of the Santa Fe Institute. UCSD faculty would visit the McKellar Plaza office regularly. I had a lunch with Montroll and Walter Kohn (future chemistry Nobel Prize winner for his density-functional theory) one day discussing the topic of $1/f$ noise and then his life story coming to the US, including being held initially as an enemy alien in an internment camp in Canada, arriving via England fleeing the Nazis takeover in his native Austria.

At lunch at the La Cote d'Azur restaurant in McKellar Plaza, one of Montroll's postdocs was trying to impress Keith Brueckner, a renowned plasma physicist at UCSD. When Brueckner showed no interest in the topic, the postdoc used his trump card saying, this is work joint with Elliott Montroll. That did catch Brueckner's attention who said something to the effect, I thought Elliott was smarter than that. I think the postdoc exaggerated Elliott's participation. Anyway, we all returned to the meal and not a physics discussion. Later, when I was interviewing for a job at the Lawrence Livermore Lab the position was to interface with Brueckner's plasma's codes against Livermore's. With no experience in plasma physics, I instead continued as Montroll's postdoc through LJI.

I spent the summer of 1976 at LJI and my first job was to write the final report on an Office of Naval Research grant. I asked, why me, as I had no contact with the work. The answer was the grant was from the ONR and it wasn't being renewed and they didn't want to waste the time of a more senior person on an action without consequence. Years later I went to work for ONR and carefully read final reports. LJI had a primitive FAX machine that was very slow and had a distinct smell and it was used to send my report to somebody.

Montroll's former postdoc, Bruce West, was a senior scientist at LJI. Back in Rochester, one day Montroll showed me a review article that he had written with Bruce that covered a number of topics from my PhD thesis. He said the paper was used as a report for a grant that he shared with Bruce through LJI. He implied that I could have been a co-author, otherwise, but I was not supported by that grant. He said he would do the next project with me. That was an article on random walks for the Studies in Statistical Mechanics series, where he was an editor. As we, together with Harvey Scher at Xerox, had brought Lévy distributions to life in our treatment of fractal time transport in disordered solids, and in the analysis of Lévy flights, we decided to trace back how Lévy probabilities came into being. This ended up as a larger report on the history of probability theory. Montroll kept asking me to write more and more and we eventually finished in 1983. It turned out the contract paid us 6 cents per word for the article that reached 121 pages. Rather than trying for a scholarly study that would pass muster with historians, we tried to find examples and insights that physicists would appreciate.

Sorry for the long introduction, but that's how Montroll's and my monograph (*The Wonderful World of Random Walks*, Studies in Statistical Mechanics, Vol. XI, North-Holland, 1984), on the history of probability came about. Due to Montroll, probability history has served me well, as I became a popular after-dinner speaker. That was a bit unusual for someone at the start of their career but, it turned out amusing and interesting vignettes about probability found receptive audiences. Now onto probability stories, but first, one last story about LJI.

Eventually an esteemed, ever growing Advisory Board for Physical Dynamics/LJI was created with winter sojourns in La Jolla, but those expenses outgrew the resources of the company causing the raising of overhead rates on contract proposals. A major multi-year grant with ONR on fluid dynamics was not renewed. While initially the company flourished, these among other factors led to its eventual move from the high rent McKellar Plaza and then LJI's gradual disappearance. At one point in the 1970s, the company was offered to buy McKellar Plaza. While it would have been a financial stretch, most of the partners were in favor, but in the end the company balked. Shortly afterwards, there was a real estate boom and the deal would have been repaid many times over.

6.2. Sticks and Bones: Luck or the Persistence of Statistical Ratios

The heel bones of certain animals, for example sheep, when tossed could land on four sides, but not the two other sides which are rounded. The bones are called astragali and they are found in piles at archeological sites. One way to toss them is to place them on your knuckles and raise your fist to throw them, hence the name knucklebones for games with astragali. They were probably used as a pastime or a way to seek guidance from the gods. They were the forerunner of dice. Perhaps as they only land on four sides they were also the forerunner of the four-sided dreidel which is spun and not thrown. The Greeks played with these bones and labeled the four reachable sides as 1, 3, 4, and 6. Throwing five bones at a time, they labeled the possible outcomes. A Zeus throw was 1, 3, 3, 4, 4. The Romans played with four bones and a 1, 3, 4, 6 was a Venus throw, while rolling four ones was called going to the dogs, a bad throw. Today, we call a roll of two ones, snake eyes and it's a bad luck throw. Playing with bones did not lead to a theory of probability as no two bones were the same. Each bone had its own unique probabilities for landing on each face. Eventually, the modern type of six-sided dice (and loaded dice) were constructed from a variety of materials and were used in gambling and divination. There is a Greek vase, circa 500 BC, illustrating a scene from the Iliad of a dice game between Achilles and Ajax during the siege of Troy, with Achilles throwing a higher number. Between them stands the goddess Athena illustrating that fate is in the hands of the gods and it is not random. Thus, luck or divine intervention was assumed to be at play, not probability. In literature, Caesar used the analogy of fate, as a roll of a die, when crossing the Rubicon to challenge the law of the Roman Republic, with the supposed quote, "Alea iacta est" (the die is cast). In Proverbs 16:33, the lot is cast unto the lap, but the decision is wholly from the Lord.

The Chinese I Ching did produce a selection process with equally likely outcomes with the choosing of long (L) or short (S) sticks. Three sticks were chosen at random, called a trigram, with eight possible outcomes, SSS, LSL, LLS, etc. Each trigram had a set of possible meanings. LLL could mean heaven, father, strong, force, northwest among other possibilities. SSS could mean earth, southwest, belly, yielding, among other possibilities. Two sets of trigrams would be randomly selected and matching opposites, such as, heaven and earth portended good fortune. For two trigrams there were 64 outcomes with a variety of interpretations, too many to recognize the permanence of statistical ratios. Anyway, the results were used as an oracle, not as an exercise in probabilities. Thus, with sticks and bones, luck or divine intervention was assumed to be at play. Imagining this is not a stretch, just ask any lottery player.

6.3.   Dice Games

Once dice became standard, various games made the rounds. One game in Italy was to keeping throwing three dice until a nine or ten was achieved. The house would win if a ten was thrown before a nine. This seemed a fair bet to the uninitiated, as looking at the table below it seems there are six combinations that add to 9 and six that add to 10.

Add up to 10	No. of ways	Add up to 9	No. of ways
[6, 2, 2]	3	[6, 2, 1]	6
[5, 2, 3]	6	[5, 1, 3]	6
[4, 2, 4]	3	[4, 2, 3]	6
[6, 3, 1]	6	[5, 2, 2]	3
[4, 3, 3]	3	[4, 4, 1]	3
[5, 4, 1]	6	[3, 3, 3]	1

There are actually 27 ways (permutations) to roll a 10 against 25 ways to roll a 9. The house bets the 10 and the player takes the 9 bet, giving a slight advantage to the house. The takeaway is that someone knew how to calculate odds. This problem was brought to Galileo who correctly pointed out the $6 \times 6 \times 6 = 216$ possibilities and the correct odds. Newton, who was born the year that Galileo died was brought the same problem and he also correctly provided the correct analysis. Knowledge is power and was kept confidential by the gambling houses.

6.4. Pascal–Fermat Letters and Their Influence on Huygens and Bernoulli

The next game, played in France, led to beginnings of recorded probability theory. The game was the house would throw a single die four times and bet on seeing a six. The player would throw a pair of dice 24 times and bet on seeing a pair of sixes. It seems the first probability would be 4/6 and the second would be 24/36, also equal to 4/6, so a fair game? However, one gambler, Chevalier de Mere, noted that single die choice was more likely to win. In 1654, he sent the problem to Pascal who provided the correct analysis which explained the difference between probabilities and expectations. Supposed the single die was thrown seven times, the probability of getting six would not be 7/6, that would be the expected number of 6's. For the single die, not getting a six in four throws is $(5/6)^4$, so the probability to get a 6 is $1 - (5/6)^4 \approx 0.5177$. A similar analysis for getting a pair of sixes in 24 throws gives the probability of $1 - (24/36)^{24} \approx 0.4914$, giving the house a slight advantage. Once again, someone knew their probability and kept the house advantage slight to keep bettors unaware of the tilt to the house.

Pascal sent his solution to Fermat, apparently looking for a mathematical colleague with whom to share a dialog. Fermat obliged. He agreed with Pascal's analysis and sent back other problems for discussion. This back and forth are known as the Pascal–Fermat letters. One example they discussed was a three-player game, with A needing one more win, and B and C each needing two more wins. In each round of play each player has an equal probability to get a win. If there is an ante, how should it be divided if the players agree to stop the game. There are needed at most three rounds with $3^3 = 27$ possiblities. The five winning sequences for B are: CBB, BCB, BBA, BBC, and BBB. The same number for C, so the ante would be split as A:B:C = 17:5:5, the Pascal–Fermat solution. It may seem strange to consider BBB as the game would end with just BB. Let's look at simpler case to make this clearer. What is the probability to get see a head (H) in two throws of a coin? The probability that this doesn't happens is when the sequence is TT, and that has probability of 1/4. One must count all possible outcomes HH, HT, TH, TT, so the probability to see a H is 3/4. We count the HH result even through a H is obtained on the first throw. D'Alembert treated, in Diderot's Encyclopedia, the following outcomes H, TH, and TT as being equally likely giving not seeing a head as probability 1/3. He might have been thinking more about what is the average number of throws needed to see a head.

6.5. The First Books and the First Paradox

The Pascal–Fermat letters made the rounds of the French salons where it caught the interest of Huygens during a visit to Paris. He is well known in the physics community, but is also a hero in probability. I've always been pronouncing his name incorrectly. For example, see the website http://seti.harvard.edu/unusual_s tuff/misc/huygens.htm where people are stopped on the street in Amsterdam and asked if they have ever heard of Huygens pronouncing it as I would. The responses were negative, then they would give the Dutch pronunciation and would get an immediate positive response. Huygens decided to write a book about probability, *De Ratiociniis in Aleae Ludo (On Reasoning in Games of Chance)* devoted to problems with a finite set of possibilities. One question had the answer of 24414065/282429536481 showing the 17th century preference for fractions over decimals. Another exercise was take cards from ace to ten in all four suits. Choose four cards at random and determine the probability that they are from four separate suits. The first card can be anything. The second card should be from a different suit and that gives 30 choices from the remaining 39 cards, etc. so the probability is

$$1 \times \frac{30}{39} \times \frac{20}{38} \times \frac{10}{37} = \frac{1000}{9139}$$

but as F. N. David points out there is a misprint in Huygens who gives the answer as $\frac{1000}{8139}$. I will be in good company for any errors that I make.

Huygens's 1657 work was only surpassed with the 1713 publication of Jacob Bernoulli's *Ars Conjectandi* (*The Art of Conjecturing*). It consisted of four parts: the first discussed Huygens' book, the other parts concerned permutations and combinations, games of chance, and application to social and economic phenomena. This great work took eight years after Bernoulli's death in 1705 to finally see publication. The task was left to his nephew, Nicolaus Bernoulli who needed time to master the topics, to travel, and to fend off criticism from his uncle Johann Bernoulli who called the book, prior to publication, "a monster that bears my brother's name." Apparently, their sibling rivalry lasted beyond reason. In the book, one finds the Bernoulli process for the probability of m successes in n trials with success p in each trial,

$$P_m(n) = \frac{n!}{(n-m)!m!}p^m(1-p)^{n-m}$$

One also finds the *first*, first passage time calculation. If each player has R coins and in each round of play a player can win or lose a coin, then on the average how long does a game last when one player has lost all of their coins. This is a random walk on a lattice of spacing unity with $2R+1$ sites going from $-R$ to $+R$. When one player wins all the coins, the other loses all their coins and the game ends, making the sites R and $-R$ absorbing boundaries. The average number of trials $\langle N(R) \rangle$ varies as R^2. In 1905 Einstein turned this problem around in his Brownian motion calculation $\langle R^2(T) \rangle \sim T$ for the mean squared displacement after a time T, with T replacing N, and distance R replacing number of coins.

Nicolaus went to study with the French mathematician Pierre Montmort and helped with his 1708 book, *Essai d'analyse sur le Jeux de Hazard* (*Essay on the Analysis of Games of Chance*). By the way, the Arabic word for dice is "al-zar" which aptly became the English word "hazard". The correspondence with Nickolaus was published in Montmort's 2nd edition of 1713. Nicolaus contributed the following problem, that seems simple enough, flip a coin until a head (H) appears. If this takes one flip, one coin is won, and that probability is 1/2 For two flips, two coins are won, and that probability is 1/4. If it takes N flips to see a head, win 2^{N-1} coins and that has probability 2^{-N}.

Now calculate the average winnings $\langle W \rangle$,

$$\langle W \rangle = \sum_{N=1}^{\infty} \frac{2^{N-1}}{2^N} = \sum_{N=1}^{\infty} \frac{1}{2} = \infty$$

The question is, what is the fair ante to play this game? The banker wants the player to ante an infinite amount as that is the player's calculated average winnings. The player counters that half of the time the winnings would only be one coin, 3/4 of the time the winnings would only be two coins, etc. Here is a taste of a fractal process with winnings increasing by an order of magnitude with that probability decreasing by an order of magnitude, generating a distribution with a mean of infinity. It is an example of a process that does not possess a characteristic size (winnings). This is called the St. Petersburg Paradox as it was discussed by another Bernoulli, (Daniel, the son of Johann) in the Commentaries of the St. Petersburg Academy of Sciences.

Probability already had a negative reputation from its connection to gambling, (an activity frowned upon or outlawed by church and state) and the St. Petersburg paradox when confronting an infinity of outcomes even had mathematicians considering it as a disreputable discipline. Mathematicians were quite happy with probability problems with a finite set of possibilities, but feared that problems with an infinite number of outcomes would not be a proper mathematical topic. It was not until the 1930s, when Kolmogorov placed probability on a formal basis that it was finally treated as a proper branch of mathematics, but as a topic in measure theory.

The next great book was Abraham DeMoivre's 1718, *The Doctrine of Chances: Or a Method of Calculating the Probability of Events of Play.* The name stuck and Thomas Bayes of Bayes Theorem entitled his book *An Essay Towards Solving a Problem in the Doctrine of Chances.* DeMoivre was imprisoned in 1685 in France when the Edit of Nantes was revoked, removing protections from Huguenots. Upon his release three years later he fled from France to England. As a young man of twenty in a foreign land, how he became a world class mathematician is impressive and a mystery to me, although he did study mathematics at the Sorbonne. He made his living by tutoring the sons of nobles and at one house came into contact with Newton's Principia. He purchased his own copy and eventually became a friend of Newton. He set himself up at Slaughters Coffee House where he calculated odds for gamblers and even had Newton as a friend. In addition to his probability book, he published the first book on actuarial science concerned with insurance rates for shipping, *Annuities Upon Lives.* Despite all of his successes, as a foreigner, he was never able to find a university position in England.

In *The Doctrine of Chances*, besides determining the odds in popular dice and card games, he derived the expansion for $\log(n-1)! \sim (n - \frac{1}{2}) - n + \log B +$ terms involving the Bernoulli numbers. With some help from James Stirling who showed $B = \sqrt{2\pi}$, the formula is now known as the Stirling approximation. DeMoivre needed a factorial expansion to show how the Bernoulli process became a Gaussian in the limit of many trials. Of course, he did not call it the Gaussian, but described the exponential as a hyperbolic logarithm. Gauss later, in 1809, showed a more general approach in his theory of errors. Robert Adrain in US also derived the Gaussian in 2D as an error distribution in surveying. DeMoivre also introduced generating functions to solve difference equations and is best known for his formula

$$(\cos\theta + i\sin\theta)^n = \cos(n\theta) + i\sin(n\theta)$$

6.6. Probability and Smallpox Inoculations

Smallpox preventive treatments has been said to originate in China with wearing of clothes of a deceased smallpox victim as a type of inoculation. Lady Montague, wife of the English ambassador to the Ottoman Empire became familiar with the practice of variolation against smallpox (Variola is latin for a mark on the skin). She herself had suffered from an episode of smallpox and had a keen interest in its treatment and prevention. Variolation was an inoculation of smallpox, cut into the skin of a person, from a smallpox victim who survived. The idea was a small dose could afford immunity, although it was not known what that dosage should be or how much was immunity provided with this method. There were casualties from variolation. Although there were earlier reports of this technique reaching England, it was Lady Montague who advocated for this treatment to infer immunity to this disease. Back in England, she had her daughter inoculated in the presence of royal physicians. The practice caught on with European royalty and spread further. In 1757, Edward Jenner was inoculated as a child and he later became a physician. As the story goes, he became acquainted with the tale that milkmaids were beautiful, meaning that they did not have the pockmarked face of smallpox victims. This led, in 1796, to his successful vaccination method of using cowpox (*vacca* is Latin for cow) to create immunity to smallpox. The name vaccine has stuck.

Daniel Bernoulli and Jean D'Alembert argued mathematically about the safety of smallpox inoculations, and even today vaccinations are still a topic of discussion. These guys stuck to the data, not hyperbole. Bernoulli, as a father of risk-benefit analysis, developed a model for life expectancy and the trade-off between receiving or avoiding inoculation. The model account for the difference in risk and gain as a function of age. Who would benefit more or risk more from an inoculation, a child or an old person. An outbreak of smallpox in Boston had a 7% mortality rate. D'Alembert had his own model and an estimate of $1/2\%$ of people would die from the inoculation. He posed the question, if there were 200 fatal diseases and 200 inoculations, who would be foolish enough to take the 200 cures when the mortality would be $1/2\%$ from each treatment. D'Alembert held sway in France, while Bernoulli advocated for the opposite. Inoculation was more prevalent in England than in France. Some claim George Washington's decision to inoculate the Revolutionary Army against smallpox was a decisive factor in winning the war of Independence, as that plague was rampant during those times. Elliott Montroll wrote about comparing vaccines to other risks, "No modern test for the characterization of a dangerous substance would allow gasoline to pass".

6.7. Bayes Theorem

Another topic in probability involves hypothesis testing and the use of Bayesian statistics. Bayes' work *Essay Towards Solving a Problem in the Doctrine of Chances* was published in 1764, after his passing in 1761. In those days, publishing was slow enough that it was not shocking that authors never saw their own books. The simplest introduction to his theorem can be presented geometrically. Consider two partially overlapping circles, A and B and their intersection written as $A \cap B$.

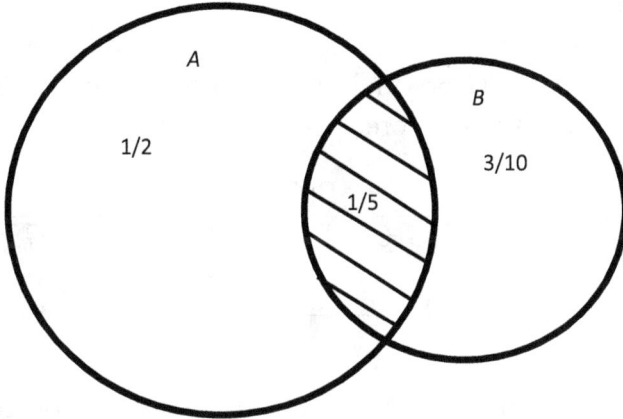

A has area $7/10$ and B has area $1/2$. Their intersection has area $1/5$. Pick a point at random and the probability that it is in the intersection A and B can be written in two equivalent ways,

$$P(A \cap B) = P(A|B)P(B) = P(B|A)P(A)$$

where $P(A|B)$ means the probability that the chosen point is in circle A given the fact that the point is also in B. Rearranging gives Bayes' formula and inserting the numbers from the figure,

$$P(A|B) = \frac{P(B|A)P(A)}{P(B)} = \frac{(2/7)(7/10)}{(5/10)} = \frac{2}{5}$$

where we noted that the area of A is $7/10$ and the area of the AB intersection is $1/5$, so $P(B|A) = \left(\frac{1}{5}\right) / \left(\frac{7}{10}\right) = 2/7$.

As an example, Bayes' theorem can solve a problem that was discussed in the Talmud. There are three draws, one has two gold coins (GG), one has two silver coins (SS), and one has a gold and a silver coin (GS). A draw and a coin are picked at random and the coin is G. What is the probability that the other coin in the drawer is also G. An incorrect line of reasoning says there are now two choices left for the contents of the chosen draw, it must have been GG or GS leaving a probability of $1/2$ that it is a GG draw. Bayes theorem finds the correct result,

$$P(GG|G) = \frac{P(G|GG)P(GG)}{P(G|GG)P(GG) + P(G|GS)P(GS)}$$

$$= \frac{1 * \frac{1}{3}}{1 * \frac{1}{3} + \frac{1}{2} * \frac{1}{3}} = \frac{2}{3}$$

Bayes theorem shows that of the two possibilities of the draws being GG and GS, given that a G was picked that the draw GG is more likely, in fact twice as likely. One can of course do the experiment to confirm this, but perhaps the Talmud was more biased towards discussion than experiment. It's more evident if the draws had $100G$, $100S$, and $1G$ $99S$. Then picking one coin and it's a G it highly likely it came from the $100G$ draw.

Bayes theorem has an aspect of confidence (or inference). Suppose you flip a coin and it comes up tails N times. How would you estimate the fairness of the coin? If you believe the coin is fair you would say the probability of getting a tail on the next coin toss is $1/2$. If you are a Bayesian you would estimate the unfairness of the coin with the probability of a tail labeled as "q" through the equation,

$$\text{prob}(q|N \text{ tails}) = \frac{\text{prob}(N \text{ tails}|q)p(q)}{\int_0^1 \text{prob}(N \text{ tails}|q)p(q)dq}$$

$$= \frac{q^N p(q)}{\int_0^1 q^N p(q)dq}$$

with the mean estimate for q being,

$$\langle q \rangle = \int_0^1 q \; \mathrm{prob}(q|N \text{ tails})dq = \frac{\int_0^1 q^{N+1}p(q)}{\int_0^1 q^N p(q) \, dq}$$

$$= \frac{N+1}{N+2}$$

where the assumption that $p(q)$ is uniformly distributed was made. Before the first coin toss $N = 0$ and $\langle q \rangle = 1/2$. After the first tail, $N = 1$ and $\langle q \rangle = 2/3$. You have evidence after one toss that a tail can occur and this as a Bayesian would have you bet that the coin is biased. As N grows $\langle q \rangle$ approaches unity.

The next step is to apply Bayes theorem to hypothesis testing, so given experimental data B what is the probability that Hypothesis A is correct, calculated in terms of the probability of finding the data result B assuming Hypothesis H is valid. This gets us into the field of statistics and inference about which whole books are available, and beyond my expertise. When I sat on NIMH study panels, much of the proposal evaluation had a focus on what was the null hypothesis, and alternative hypotheses, are all factors effecting the experimental outcome known, and if the proposed experiments could make a determination. Criteria included if too many lab animals would be sacrificed, or not enough, depending on what method of hypothesis testing was employed. This was very different from the panels I've sat on deciding on physics proposals which have objectives, and usually not hypotheses.

6.8. Probability Moves to France

While we might disagree with D'Alembert's defense against inoculation and his incorrect calculation of coin flipping probabilities, he did end up in his study of waves having the $\frac{\partial^2}{\partial t^2} - \nabla^2$ operator named after him. This shows up in general relativity as $\partial^\mu \partial_\mu$. He had a young protégé, Nicolas Condorcet, who sought to apply probability mathematics to social and economic topics. One problem was determining the optimal size for a jury assuming a probability for a juror reaching the correct verdict. His model favored 15 jurors and a plurality of 5 to free or convict. Condorcet played important roles in the French Revolution including plans for universal free education. Ultimately, he ran afoul of the most radical elements with his liberal ideas, and was arrested and died in prison.

Condorcet's work did have an influence on Laplace, although not explicitly cited his book, *Theorie Analytique des Probabilites*. Todhunter's 1865, *The History of the Mathematical Theory of Probability*, devoted about one quarter of his book to Laplace's work. Laplace's book contained much from previous authors, and his style was not to dwell on that fact. His work introduced new ideas on generating functions leading up to the important now extensively used Laplace transform to solve difference equations and linear differential equations. We will use it to solve continuous time random walk equations. Laplace solved many integrals that arise in probability, and are today treated as standard in calculus books, including

$$\int_0^\infty \exp(-x^2)dx = \frac{\pi}{2}$$

and

$$\int_0^\infty \cos(rx)\exp(-a^2x^2)dx = \sqrt{\frac{\pi}{4a^2}}\exp\left(-\frac{r^2}{4a^2}\right)$$

showing Gaussians transform into Gaussians. He was able to recast many problems in terms of picking red and white tickets out of an urn and then determining the probability of the next picks. Here's an example, if an urn has an infinite number of red and white tickets and p white tickets are picked along with q red ones, then the probability P of picking more tickets that m will be white and n will be red is,

$$P = \frac{\int_0^1 x^{p+m}(1-x)^{q+n}dx}{\int_0^1 x^p(1-x)^q dx}$$

Laplace was the man who put the integral into probability. He taught at École Polytechnique and had Poisson as a student, his favorite student. Many years latter Mandelbrot graduated from École Polytechnique and I had the pleasure to join him at the French Embassy in Washington DC for the Ecole's 200th anniversary. He spoke about fractals in a pre-dinner talk, and the highlight of the meeting seemed to be the pate de foie gras. Poisson replaced Fourier on the faculty when the latter accompanied Napoleon to Egypt. Poisson is known for many accomplishments, but here we go back to the Bernoulli trials where DeMoivre showed in the limit of a large number of trials, n, the result approached a Gaussian. Poisson took a different limit of the probability of success, p, going to zero, that is success is a rare event. But now, he also lets n the number of trials also go to infinity and keeping the limit $np \to \lambda$ Explicitly

$$P_m(n) = \text{probability of } m \text{ success in } n \text{ trials}$$

$$= \frac{n!}{(n-m)!m!}p^m(1-p)^{n-m}$$

$$= \frac{n(n-1)\cdots(n-m+1)}{m!}p^m(1-p)^{n-m}$$

as $n \to \infty, p \to 0, np \to \lambda$, then

$$P_m(n) \sim \frac{n^m p^m \exp(-np)}{m!}$$
$$= \frac{\lambda^m \exp(-\lambda)}{m!}$$

Voila, the Poisson distribution. Note, that even although this started as a rare event exercise, that λ need not be small, as the decay of radioactive elements can attest. Poisson's work was basically ignored by his French contemporaries as he attempted to apply probability to criminal and civil matters. His 1837 book *Recherches sur la probabilite des jugements en matiere criminelle et en matiere civile* was not well received. The first published application of Poisson's Law was in Germany in 1894 for the analysis of twenty years of data of the rare event of cavalry riders being kicked to death by their horse. A memorable use of Poisson's Law was by William Gosset a Guinness employee in Dublin. He counted yeast cells in samples to ensure a proper fermentation and found the data was well fit by a Poisson distribution. Gosset published this in 1904 and other works on probability and statistics under the pseudonym "Student" as he believed Guinness would not approve publicizing their brewing methods. Today, Gosset's Student t-distribution is well known for assessing statistical significance.

6.9.   Probability Moves to Russia

As a student, Pafnuty Chebyshev came across Poisson's book and was not repulsed by Poisson's social topics. Even Jakob Bernoulli considered such topics in his Ars Conjectandi. In 1846 Chebyshev submitted his Master's thesis at Moscow University entitled, "An Essay on an Elementary Analysis of the Theory of Probability". He is well known for the Chebyshev inequality,

$$P(|X| \geq t) = \int_{|x|=t}^{\infty} p(x)dx \leq \int_{|x|=t}^{\infty} \frac{x^2}{t^2} p(x)dx$$

$$= \frac{1}{t^2} \int_{|x|=t}^{\infty} x^2 p(x)dx \leq \frac{1}{t^2} \langle x^2 \rangle$$

Lecturing at the St. Petersburg University, he had two exceptional students, Andrey Markov and Aleksandr Lyapunov. Markov invented Markov chains (more about that later), and Lyapunov who is best known for his work on dynamical systems and the Lyapunov exponents now used in chaos theory, and he also used characteristic function methods to prove the Central Limit Theorem under new wider conditions.

6.10. Bertrand's Paradox

Back in France, Joseph Bertrand in 1879 wrote *Calcul de probabilities*, then Henri Poincare in 1896 wrote, *Calcul de probabilities*. Back in St. Petersburg, Markov, in 1900, wrote *Ischislenie Veroyatnostei* which is translated as *Calculus of Probabilities*, and then in 1925, Paul Lévy, wrote *Calcul de probabilities*. So not much originality in book titles. Lévy did write other books. In 1937 he wrote *Theorie de l'laddition des variables aleatoires* (*Theory of the Addition of Random Variables*) and in 1948 *Processees stochastiques et movement brownien* (*Stochastic Processes and Brownian Motion*).

In probability, there is no magic formula like the Newton, Maxwell or Schrödinger equations that one can plug into. All you get to start is that probabilities sum to one and all probabilities are positive or zero. One must carefully analyze each problem to set it up correctly and it's easy to go wrong. Bertrand's example is just one example of how the wording can be ambiguous and the care one must exercise in probability. We'll end here with Bertrand's paradox.

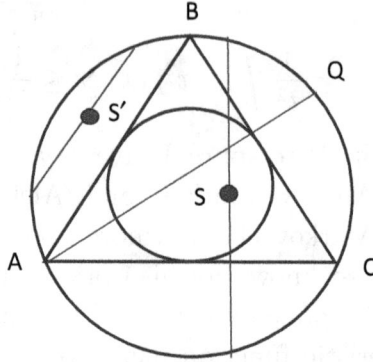

Inscribe an equilateral triangle inside a circle and ask, if choosing a chord at random what is the probability that it is longer than the side of the triangle? There is an obvious answer of 1/3. Pick a point on the circumference and label it A. Draw the equilateral triangle with a vertex at A and label the other vertices as B and C. The arc between B and C is 1/3 of the circumference. Now pick at random a second point on the circumference and label it Q. If Q lies between B and C, the chord AQ will be longer than the side of the triangle.

Another approach gives and answer of 1/4. What is going on? Look at the circle inscribed inside the triangle of area 1/4th of the larger circle and use a different method of drawing a chord by choosing, at random, a point inside the larger circle. Label it as S and use that point as the midpoint of a chord. If S falls inside the smaller circle the chord using S as a midpoint will be longer than the side of the triangle. Pick the point S' outside of the inscribed circle and use it as the midpoint of a chord. That chord will be shorter than the side of the triangle. Using a two-point random selection linear measure (the circumference) one gets 1/3, and choosing a single point, using an area measure one gets 1/4. The answer depends on how you randomly select a chord and decide what is uniformly distributed.

I stop here in the late 1800s, as the field of probability and statistics blossomed in the 20th century and several volumes would be needed to track it all. In the rest of this essay I only focus on the aspects of random walks.

CHAPTER 7

Lévy Flights and the Random Walks of Weierstrass and Riemann: An Early Run-in With Fractals

7.1. Benoit Mandelbrot's 1972 Colloquium

While neither Weierstrass nor Riemann worked on random walks, we have named two walks in their honor as they rely on famous functions studied by these men.

The story starts as usual with my thesis advisor, Elliott Montroll who in the early 1960's was the Director of General Sciences at IBM Yorktown. This overlapped at IBM with the young Benoit Mandelbrot, later known as the father of fractals. In 1972, when Mandelbrot gave the physics colloquium at the University of Rochester, Montroll, for the only time, recommended his graduate students to attend the colloquium. Leo Kadanoff had a rule of when one could skip talks, (1) if the speaker wore a tie, (2) if the speaker did not wear a tie. I decided to go to Mandelbrot's lecture. I think he wore a tie.

The lecture was strange to say the least. Mandelbrot showed a picture of dots and said they looked like star clusters.

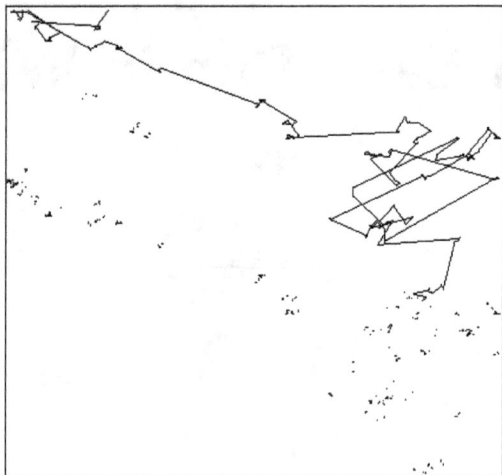

The black dots in the picture above (supposedly look like stars in the night sky) are an example of the pictures that Mandelbrot was showing. The trajectory above has those dots as turning points projected below. This is called a Lévy flight.

For one of the pictures, Mandelbrot went on to say that the dots had a dimension of around 1.7. He didn't show the trajectories, just the turning points. Hands went up and the question was asked, what is a dimension that is not one or two or three? Mandelbrot then put up other pictures of dots with denser sets saying they have a dimension larger than 1.7 and thinner sets having lower fractional dimensions. He would not define what is a fractional dimension and just asked the audience to use their imagination. Why was he comparing these pictures to star clusters? That didn't make any sense. Was there any physics involved and what was the math behind it all? The talk did not go over well and annoyed the audience of physicists. But I was fired up. After the talk, I asked Montroll if he could explain what it was all about. He said the points were drawn by a random walk taking a number of small steps, forming a cluster of points visited, followed by a long jump, then again a cluster of small steps, and after some of these clusters were formed then an even larger step was taken and eventually clusters within clusters within clusters were formed ad infinitum. Montroll mentioned this was related to work of Paul Lévy on random variables whose moments were infinite. He said Lévy's work was beautiful mathematics, so it should one day find its way into physics, but he did not know where it would play a role. Of course, that would change in a dramatic fashion when the field of fractals burst onto the scene and the scale invariance of fractals melded with the physics of topics like the renormalization group for percolation and phase transitions.

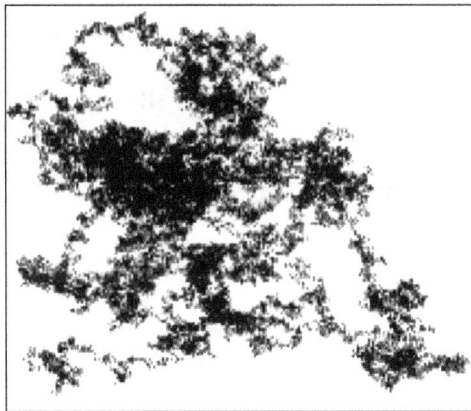

The picture shows a random walk in 2D taking steps of unit length randomly.

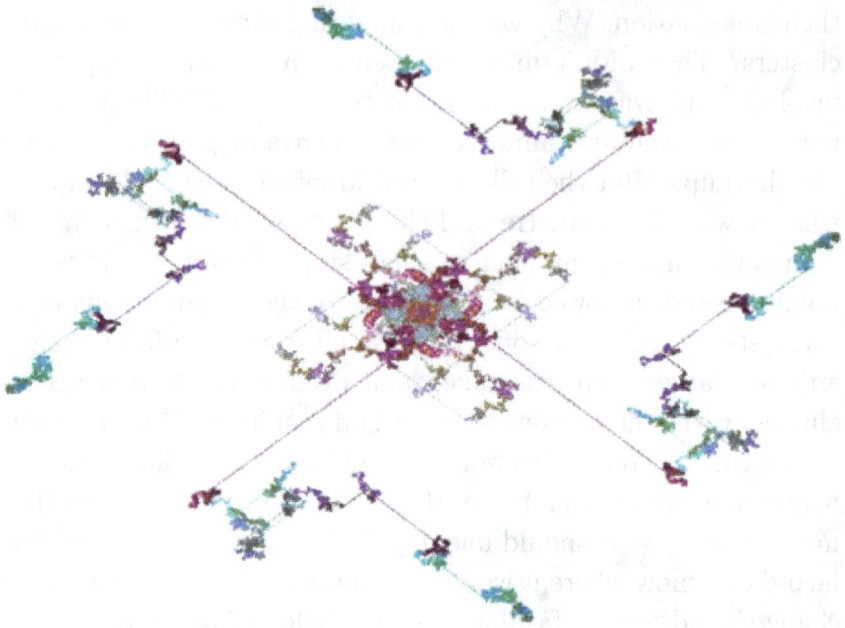

The picture shows a trajectory of a 4-fold symmetric Zaslavsky map. Each leg looks like a Lévy flight, but this is from a nonlinear mapping. More about this in the last chapter.

7.2.   The Weierstrass Random Walk

After the Mandelbrot colloquium, I went back to my office and immediately wrote down the probability for a random jump of displacement x,

$$p(x) = \frac{n-1}{2n} \sum_{j=0}^{\infty} \frac{1}{n^j} (\delta_{x,b^j} + \delta_{x,-b^j}) \qquad (7.1)$$

where b and n are constants both greater than one. This probability has jumps of an order of magnitude longer in base b that occur but with an order of magnitude less probability in base n. This can easily be generalized to include jumps in the y and

z directions. It implies that most likely about n jump of length one happen forming a cluster, then a jump of length b occurs to start a new cluster. This repeats around n times forming n clusters separated by a distance b, before a of jump of length b^2, and so on over all scales in length and probability of occurrence generating clusters of clusters within clusters ad infinitum, i.e., a fractal pattern. If a characteristic size existed, say the mean square displacement, then there would be a scale and no scale invariant behavior would exist, so infinite moments are required for scale invariance. To get the fractal dimension of the sites visited we Fourier transform equation (7.1), $x \to k$ to go from $p(x)$ to $p(k)$. Normally, one would to have $p(k)$ with a star or tilde to distinguish it from $p(x)$ as they are different functions, but I hope the meaning is clear when omitting the star or tilde. We have the Fourier transform,

$$p(k) = \sum_{x=-\infty}^{\infty} \exp(ikx)p(x)$$

$$= \frac{n-1}{n} \sum_{j=0}^{\infty} n^{-j}\cos(kb^j) \tag{7.2}$$

One can see that $p(k)$ satisfies the scaling equation

$$p(k) = \frac{1}{n}p(bk) + \frac{n-1}{n}\cos(k) \tag{7.3}$$

and the homogeneous part of the equation has a solution of k^β with $\beta = \frac{\ln(n)}{\ln(b)}$, an exponent that's not an integer and wouldn't appear in a Taylor series. When $b^2 > n$, then $\beta < 2$, making $\langle x^2 \rangle = \infty$. My model to produce the Lévy flight "star clusters" of Mandelbrot is

$$p(x,y) = \frac{n-1}{4n} \sum_{j=0}^{\infty} \frac{1}{n^j}(\delta_{x,b^j} + \delta_{x,-b^j}) + \frac{n-1}{4n} \sum_{j=0}^{\infty} \frac{1}{n^j}(\delta_{y,b^j} + \delta_{y,-b^j})$$

$$\tag{7.4}$$

Although I wrote this in 1972, the complete analysis had to wait until 1981 for Barry Hughes to join Montroll as a postdoc. Barry was a student of Barry Ninham in Australia. Ninham was a former Montroll graduate student and the co-author, with Michael Barber of the 1970 book, *Random and Restricted Walks*, that reviews a fair bit of Montroll's work on random walks. Hughes joined Montroll at the University of Rochester, but Montroll had turned 65 and was in the process of moving to the University of Maryland. I had joined Montroll in Maryland in August 1980 when he remarked that he had a postdoc in Rochester, so I suggested why not have him move to Maryland. It turned out that Barry was a master of mathematical analysis and when I showed him Eq. (7.2) he immediately said, that's the Weierstrass equation for a continuous, but non-differential curve. Weierstrass met the challenge of coming up with a curve that was continuous and nowhere differentiable. There was no application, just a warning that one's intuition about what is possible can be wrong. I had never heard of the Weierstrass function, but I decided to call this the Weierstrass random walk in his honor. Here's what Barry did next, something I knew nothing about, he wrote the cosine in Eq. (7.2) in terms of its own Mellin transform,

$$\cos(k) = \frac{1}{2\pi i} \oint \frac{\Gamma(s) \cos\left(\frac{\pi s}{2}\right)}{|k|^s} ds, \, Res > 0 \qquad (7.5)$$

where the gamma function has simple poles at 0 and the negative integers and the cosine has zeros at the even integers, so the residue theorem picks up terms of k in even powers and reproduces the series for the cosine.

Substituting the inverse Mellin transform for the cosine into Eq. (7.2) to give

$$p(k) = \frac{1}{2\pi i} \frac{n-1}{n} \sum_j n^{-j} \oint \frac{\Gamma(s) \cos(\pi s/2)}{|kb^j|^s} ds$$

and performing the sum and switching it with the contour integral gives,

$$p(k) = \frac{1}{2\pi i} \frac{n-1}{n} \oint \frac{\Gamma(s)\cos(\pi s/2)}{1 - n^{-1}b^{-s}} k^{-s} ds \qquad (7.6)$$

which has a pole at $s = -\frac{\ln(n)}{\ln(b)} \pm \frac{2\pi i j}{\ln b} j = 0, \pm 1, \pm 2, \ldots$ and simple poles at $s = 0, -2, -4, -6, \ldots$. If $b^2 > n$, this gives rise to a term k^β with $\beta = \frac{\ln(n)}{\ln(b)}$, and this term dominates the k^2 term as $k \to 0$, so for small k values, and except for $s = -2$, ignoring the poles at the negative even integers, the Fourier transformed probability behaves as,

$$p(k) \sim 1 - k^\beta + O(k^2) \qquad (7.7)$$

and transforming back and using l for integer lattice distances,

$$p(l) \sim \frac{1}{2\pi} \int_{-\pi}^{\pi} \exp(-ikl)\exp(-|k|^\beta), \quad \beta < 2 \qquad (7.8)$$

for large l,

$$p(l) \sim \frac{1}{l^{1+\beta}} \qquad (7.9)$$

Ignoring all the technical details, Eq. (7.8) is more or less, the form of a Lévy distribution. When the second moment is finite, then $\beta = 2$. If one let β be greater than 2 in Eq. (7.8), it would produce negative probabilities, but the analysis for the poles always would have $\beta = 2$ as an upper bound. If one started with Eq. (7.9) with $2 < \beta < 3$, then $p(k) \sim 1 - k^2 + k^\beta$ and Gaussian behavior would asymptotically ensue.

The second k derivative of $p(k)$ diverges at $k = 0$ for $\beta < 2$, so $\langle x^2 \rangle = \infty$. The full exact formula for $p(k)$ can be derived and it is complicated and includes a factor periodic in log k with period log b, but there is no need to get into all that. If you are interested see B. D. Hughes, M. F. Shlesinger, and E. W. Montroll, *PNAS* **78**, 3287 (1981). But note that,

$$f(x) = f(\lambda x)$$

has a more general solution than $f(x) = $ constant, i.e.,

$$f(x) = \text{constant} \sum_n e^{2\pi n i \frac{\ln(x)}{\ln \lambda}}$$

What we have tied together is, that the Weierstrass function, as a random walk generating function, produces a fractal set of visited points of dimension $\frac{\ln(n)}{\ln(b)}$. This corresponds to a self-similar set of clusters over length scales in powers of b, with n being the number of clusters in the next sized cluster. So Weierstrass' work turned out to be an early step into fractals and scale invariance that shows up as generating a fractal random walk.

By the way if one does the analysis in three dimensions, but with spherically symmetric jumps,

$$p(x) = \frac{n-1}{n} \sum_{j=0}^{\infty} n^{-j} \delta(l - b^j) \tag{7.10a}$$

$$p(k) = \frac{n-1}{n} \Gamma\left(\frac{3}{2}\right) \sum_{j=0}^{\infty} \frac{1}{n^j} \left(\frac{1}{2} k b^j\right)^{-1/2} J_{1/2}(k b^j) \tag{7.10b}$$

This realizes that the Weierstrass function is a lacunary series of cosines, its generalization is a lacunary series of spherical Bessel functions. Lacunary series with increasing gaps have dense singularities on their radius of convergence preventing analytic continuation. Here we tied this into a non-Taylor expansion using Mellin transform techniques and found the set of sites visited in the random walk that Weierstrass generates to produce a Levy flight, a topic championed by Benoit Mandelbrot. Benoit's uncle, Sjolem Mandelbrojt, a renowned French mathematician, studied lacunary series and thus has a belated potential tie into fractals. He was also a founder of the Bourbaki group, a collection of mathematicians that tried to do all mathematics formally without relying on intuition. Benoit's approach on developing fractals was very much based on intuition, and he has written that one of the factors on his leaving France for the US was to get away from the domineering influence of Bourbaki in France.

In 3D, for a fractal set of dimension $d < 3$, like the turning points in a Lévy flight, the number of new sites encountered when expanding a sphere of radius R, grows as R^{d-1} more slowly than R^2. Going back to Mandelbrot's star analogy, light decreases in intensity with distance from a star as $1/R^2$. Early in career, Paul Lévy pointed out for a fractal configuration of stars the night sky would appear as it does and not with a uniform dull white background. So, fractals are a solution to Olbers paradox of why the night sky is not a uniform white. The specifics are more complicated, such as, gas cloud can absorb light, but fractals were an early application to astrophysical structures.

Let's look again at the Green's function for return to the origin $G(0, z)$, but for now for a Weierstrass-type random walk,

$$G(0, z) = \frac{1}{2\pi} \int_{-\pi}^{\pi} \frac{dk}{1 - zp(k)} \sim \frac{1}{2\pi} \int_{-\pi}^{\pi} \frac{dk}{1 - z + k^\beta} \qquad (7.11)$$

Making the change of variables to $m^2 = k^\beta$ we can rewrite Eq. (7.11) as,

$$\frac{1}{2\pi} \int_{-\pi}^{\pi} \frac{2}{\beta} \frac{m^{\frac{2}{\beta}-1} dm}{1 - z + m^2} \qquad (7.12)$$

giving the random walk an effective dimension of $2/\beta$. If $\beta \le 2$ the return to the origin in 1D will no longer be a probability of one. This is a taste of the many new types of dimensions that can be applied to fractal shapes and processes. Another is the spectral dimension of the number of new sites visited by a random walk. On a fractal structure, such as a critical percolation network, this dimension is different than the percolation dimension.

7.3. The Riemann Random Walk

Let's start with what has been called the most famous unsolved problem in mathematics: The Riemann Hypothesis. It asks for the location of the zeroes of the Riemann zeta function. That puzzle is not solved here. This is connected to the density of prime numbers, but I place it into the context of a fractal random walk.

Long ago, in 1859, Riemann wrote an eight-page paper on the density of prime numbers using innovative complex analysis. He did not solve the prime number problem, that only occurred more than three decades later. Nevertheless, Riemann's work was epoch-making due to the variety of new methods that he introduced to number theory. He studied the sum,

$$\zeta(s) = \sum_{n=1}^{\infty} \frac{1}{n^s} = 1 + \frac{1}{2^s} + \frac{1}{3^s} + \cdots$$

but with s being a complex number. At $s = 1$ this is the harmonic series which is divergent. The primes enter as we can write,

$$\zeta(s) = \prod_{p \, \text{primes}} \frac{1}{1 - p^{-s}} \tag{7.13}$$

This can be shown as

$$\frac{1}{2^s} \zeta(s) = \frac{1}{2^s} + \frac{1}{4^s} + \frac{1}{6^s} + \cdots$$

and

$$\left(1 - \frac{1}{2^s}\right) \zeta(s) = 1 + \frac{1}{3^s} + \frac{1}{5^s} + \cdots$$

with all terms with a factor of 2 are missing, continuing this analysis,

$$\left(1 - \frac{1}{3^s}\right) \left(1 - \frac{1}{2^s}\right) \zeta(s) = 1 + \frac{1}{5^s} + \frac{1}{7^s} + \cdots$$

with all terms with factors of 2 and 3 being missing. Continuing this procedure for all primes recovers Eq. (7.13). We will be interested in the zeroes of $\zeta(s)$ and these will show up as the poles of $1/\zeta(s)$ which we now study by examining

$$\frac{1}{\zeta(s)} = \prod_{p\,\text{primes}} (1 - p^{-s}) = \left(1 - \frac{1}{2^s}\right) \left(1 - \frac{1}{3^s}\right) \left(1 - \frac{1}{5^s}\right) \cdots$$

$$= \sum_{n=1}^{\infty} \frac{\mu(n)}{n^s}$$

where $\mu(n) = (-1)^m$ if n is a product of m distinct primes, and $\mu(n) = 0$ otherwise, and $\mu(n)$ is called the Mobius function.

We now consider the following random walk on a lattice with $\beta < 1$,

$$p(l) = N \sum_{n=1}^{\infty} \frac{1 \pm \mu(n)}{n^{1+\beta}} \frac{1}{2} [\delta_{l,n} + \delta_{l,-n}]$$

where N is a normalizing factor, and the $1 \pm \mu(n)$ term is either 2 or 0. The Fourier transform is

$$p(k) = N \sum_{n=1}^{\infty} \frac{1 \pm \mu(n)}{n^{1+\beta}} \cos(nk)$$

and using the inverse Mellin transform trick

$$= \frac{N}{2\pi i} \oint \sum_{n=1}^{\infty} \frac{1 \pm \mu(n)}{n^{1+\beta}} \frac{\Gamma(s)\cos(\pi s/2)}{|nk|^s} ds$$

$$= \frac{N}{2\pi i} \oint k^{-s} \Gamma(s) \cos(\pi s/2) \left[\zeta(1+\beta+s) \pm \frac{1}{\zeta(1+\beta+s)} \right] ds$$

We have simple poles at $s = 0, -2, -4, \ldots$ from the cosine gamma function terms, a simple pole at s $= -\beta$ from the zeta function, and all other poles arise from the zeroes of zeta function in the denominator. Remember that $p(k)$ represents a transformed probability and that places restrictions on where the zeta function can have zeroes. First, any zeroes need to appear as complex conjugate pairs to keep $p(k)$ real. The probability is normalized to unity and the poles from $s = -\beta$ and 2 produce terms of the form $1 - k^\beta + O(k^2)$. If there is a zero of zeta with $s > -\beta$, say, s $= -\beta + b$, then this adds a term forming $p(k) \sim 1 \pm k^{\beta-b} - k^\beta + O(k^2)$ as $k \to 0$. As $p(k)$ needs to be ≤ 1 as $k \to 0$, choosing the plus sign for $k^{\beta-b}$ would have the probability violating $p(k) > 1$ near k going to zero. This means that any zeroes must have the real part of $s < -\beta$. Additionally, Riemann derived the equation

$$\zeta(1-s)(2\pi)^s = 2\cos\left(\frac{\pi s}{2}\right)\Gamma(s)\zeta(s)$$

that related $\zeta(s)$ to $\zeta(1-s)$, and places any zeroes to real parts between $s = -1-\beta$ and $s = -\beta$. The famous Riemann Hypothesis is that all the zeros are complex and lie on the line Re $s = -1/2-\beta$. The analysis here used probability to place the zeroes in a strip of width unity, a well-known result. Perhaps, a reader can come up with other probability requirements that can better locate or put restrictions on the complex zeroes of Riemann's zeta function.

I once wrote a one-page submission to *Nature* for their science fiction page. It was about the Earth Ship Riemann seeking mathematicians in civilizations across the galaxy to discover if anyone had solved the Riemann Hypothesis. Afraid of germs, they would send down a shuttle to bring mathematicians back to their ship which had a rotating section to duplicate the gravity effect of the host planet. The earth and the other group would communicate through touch screens and start a process to learn each other's math until both sides could discuss Riemann's complex analysis. In the story, there was no success.

CHAPTER 8

A Paul Lévy Conference Menu

Since the last chapter started with Benoit Mandelbrot's colloquium on Lévy flights, I'll jump ahead 22 years to show you the dinner menu in Cap Ferrat next to Nice in France. This was for a workshop that Uriel Frisch, George Zaslavsky and myself organized in honor of Paul Lévy's work in probability. The menu shows my scribbles while listening to presentations at the dinner. The meeting was held at the Observatoire de Nice and Helene Frisch, the granddaughter of Paul Lévy, was one of the participants. The conference proceedings are in *Lévy Flights and Related Topics in Physics*, eds. M. F. Shlesinger, G. M. Zaslavsky, and U. Frisch, in *Lecture Notes in Physics*, Vol. 450 (Springer, 1995).

RESTAURANT LA COSTIERE 3,AVE.ALBERT 1
06230 ST.JEAN CAP FERRAT
TELEPHONE 93 76 03 89

VUE PANORAMIQUE

RECEPTION DU 30 JUIN 1994

LE BOWLE
(COCKTAIL DE FRUITS FRAIS MARCERES AU COGNAC,ARROSES DE VIN BLANC
ET CHAMPAGNE)

AMUSES - BOUCHE

LE SAUMON NORVEGIEN CRU CREME ANETH SUR BLINIS

LES BEIGNETS DE SCAMPIS SAUCE TARTARE SUR MESCLUN

LE BOEUF STROGANOFF
(LE FILET DE BOEUF EMINCE FLAMBE AU COGNAC EN SAUCE STROGANOFF:
TOMATES,POIVRONS ROUGES,CHAMPIGNONS ET FINES HERBES EN CREME FRAICHE)

LE BRIE DE MEAUX SUR POUSSES D'EPINARDS EN SALADE

LES FRAISES MELBA
(LES FRAISES DE CARROS,GLACE VANILLE,CREME CHANTILLY)

VOTRE MOKA

VOS VINS:

MUSCADET "LA CHATELLENIE"
BOURGOGNE DU CHAPITRE 1985 MAISON JAFFELIN
MERCUREY 1985 MAISON JAFFELIN

CHAPTER 9

Elliott Montroll: An Appreciation

9.1. Elliott Montroll in his 20's

It was the Fall of 1970 and I was starting graduate school at the University of Rochester in upstate New York close to Lake Ontario. In my acceptance package, the university brochure listed three types of housing, a graduate living center, and locations for townhouses that seemed to be for families. The graduate living center seemed ideal as it was walking distance to campus, so that's what I choose. Well I got to the university to discover that there was no vacancy at any of the graduate housing and all I could do was to put my name on a waiting list. I ended up finding an apartment to share with two physics grad students in not the best part of town.

At Stony Brook I was a math major until my last year when I took enough physics courses to also qualify for a physics degree. I ended up asking a Hungarian math professor who taught complex analysis for a letter of recommendation. I have no idea if he ever wrote one. Nevertheless, based on my GRE score and math and physics transcripts and my two years working at the

Stony Brook Van der Graaf lab, I was accepted to Rochester. It turned out that Elliott Montroll won a large NSF grant for a center that he called the Institute for Fundamental Studies. Montroll through his DOD connections also managed the large omnibus grant for the Center for Naval Analyses. These awards brought in funds to support new graduate students. My class was 13 students. About half were foreign and they seemed better prepared than the American students. Why such a low number. It turned out in 1968 when funds were plentiful Rochester accepted 48 students and in 1970 the funds only allowed for a small class, in part even that number was thanks to Montroll's grants.

At the beginning of the year I knocked on Montroll's door and there was no answer, so I knocked louder and finally heard a come in. It turned out that I should have entered through the door to his secretary's office. Nevertheless, he was friendly. I started by saying that Max Dresden suggested that I work with him. He asked me, with interest, what math courses I had taken and said come back at the end of the second semester to talk about joining his group. It was rather intimidating in that he had nine grad students and ten postdocs plus three other professors associated with him.

I came back at the end of the second semester and asked for financial support over the summer. In those days, grad students got $300 a month. By this time, I moved to the graduate living center where my rent for a shared two-bedroom apartment was $75. I lived comfortably on that and thought I could live royally on eventually becoming a postdoc as they made $10,000 a year. How naïve. Anyway Montroll pulled out five papers from his filing cabinet and said see if, any of these, interest you. He said I should read the collection of famous papers in Nelson Wax's book of collected papers *Noise and Stochastic Processes* and that he was going away for the summer and would be back sometime in September, and that was that. I had the summer free to read and think while my fellow grad students worked long hours in labs. I spent a glorious summer reading Wax, the two classic volumes of Feller's *An Introduction to Probability* and the pure mathematics book by Spitzer *Principles of Random Walk*. But it was the Montroll–Weiss paper of 1965 called Random Walks on Lattices II that really was fun to read. The Spitzer math book gave the impression that to make any advance would be hard and needing a tedious proof. Much later I wrote a short paper on the stretched exponential distribution from a probability standpoint. Afterwards, Joel Lebowitz at Rutgers University wrote a long paper on the stretched exponential. I asked him why his paper was so long when mine was so short. He said my paper was an upper bound and it took much more space to also prove it was also lower bound (or maybe it was the other way around). I preferred the Montroll style.

The Montroll–Weiss paper was so clear that I felt I could start my own research. The paper used a Green's function approach that was clear to apply to solve a wide array of random walk questions, several of these can be found in Chapter 1. As an undergrad in the math physics course, the study of Green's functions was overwhelming and went by quickly and here with Montroll it seemed intuitive and easy to use and exploit. Sometime later the work was referred as the Continuous Time Random Walk (CTRW) and a special volume of collected papers *The CTRW Still Trendy: Fifty Year History, Current State and Outlook* covering many topics was published in 2018 in the *European Physics Journal B*. I wrote for that publication "At the 50-year point the Montroll–Weiss CTRW has turned out to create a family of colleagues and friends that has spanned disciplines, continents, and generations." So, the Montroll–Weiss paper has held up for more than a half century and still inspires many works.

So, who was Elliott Montroll?

Elliott was a chemistry undergrad who switched to math for the PhD and had an amazing start in his youth that he maintained throughout his career. He told me that he was losing his sense of smell in chemistry labs and so for grad school, at the University of Pittsburgh, he switched to math in 1937, a topic in which he excelled. He was awarded his PhD in 1939. The theory of imperfect gases was a hot topic in those days and the leading figure was Joseph Mayer at Columbia University. It happened that Gregory Wannier, then at Pittsburgh, asked Montroll to give a seminar lecture on Mayer's work. Wannier was a Swiss physicist who achieved later fame with the creation of Wannier functions and Wannier excitons in solid-state physics. The problem was to calculate the partition function Z for an imperfect gas to go beyond the well-known ideal gas law. In general, for N particles interacting through a potential,

$$Z = \frac{1}{N!} \int \cdots \int \exp\left[-\frac{\sum_{i<j} \varphi(r_{ij})}{kT}\right] dr$$

where r is an N-dimensional vector. Writing in terms of Mayer f functions,

$$\exp(-\varphi(r_{ij})) \equiv 1 + f_{ij}$$

then,

$$\exp\left[-\frac{\sum_{i<j} \varphi(r_{ij})}{kT}\right] = 1 + \sum_{i<j} f_{ij} + \sum_{i<j}\sum_{k<l} f_{ij}f_{kl} + \cdots$$

What a complicated equation. Well, Montroll went from configuration space r to Fourier space k to calculate ring integrals,

$$\int \cdots \int f_{12} f_{23} \cdots f_{n,n+1} f_{n+1,1} dr_1 \cdots r_{n+1}$$

This was a new idea and approach and these rings for sizes up to an N could be calculated and summed. This became Montroll's PhD thesis and he went to work with Joe Mayer at Columbia University. His method was also applied to the study of the Ising model, a phase transition model of a ferromagnet. The method is now called the transfer matrix approach. Montroll could solve the 1D model where there was no phase transition and in 2D he could find high and low expansions in temperature. Wannier brought Montroll's method to Hendrik Kramers in the Netherlands one of the leading theoretical physicists in Europe. They produced the famous Kramers–Wannier paper showing how to find the critical temperature of the 2D Ising model. They write "The elegant form of procedure used here is due to Mr. E. Montroll who applied it first to the theory of molecular chains". This is remarkable! The standard course for an American physicist was to move to Europe to apprentice with the top physicists. Here, instead a top famous Dutch physicist is thanking an American graduate student.

The work with Mayer is now known as the Mayer–Montroll equations. After Columbia, Montroll went to Yale to work with Lars Onsager. For Montroll's 60th Birthday celebration, Onsager wrote,

I got a letter from Joe Mayer recommending a young fellow, indeed recommending him very highly. The young fellow had worked with Wannier before and spent a year with Joe and Maria. On the strength of that recommendation, I used the strongest pressure I had at my disposal to wedge an opening and Elliott came to occupy it. I was very glad that I had made the effort. And incidentally, so is everyone else in the Chemistry Department at Yale. I think everyone found Elliott very interesting to talk to. He brought the news of this development on the Ising model that Wannier had brought home after working with Kramers.

Montroll taught Onsager the matrix method to solve the Ising model and Onsager succeeded to solve for the first time the partition function for the 2D Ising model. That achievement was noted in his selection for the Chemistry Nobel prize. And by the way, the Maria in Joe and Maria, was Maria Goeppert Mayer, herself a future Nobel Prize winner for work in nuclear physics. Also by the way, Montroll tells the story that sometimes Onsager would start laughing in conversations for no apparent reason. When questioned he admitted, some words and cadences in English sounded like dirty limericks in Norwegian.

To give the flavor of Montroll's idea, consider spins that can be up or down on a line. The partition function Z involves a sum over all spins configurations 1 to N, but Montroll put the spins on a ring, so spin N was the neighbor of spin 1. The partition function now has the terms,

$$Z = \sum_{\text{all spins}} e^{-\frac{E(\sigma_1, \sigma_2)}{kT}} \cdots e^{-\frac{E(\sigma_N, \sigma_1)}{kT}}$$

and defining a transfer matrix $t_{ij} = e^{-\frac{E(\sigma_i, \sigma_j)}{kT}}$, then

$$Z = \sum_{\text{all spins}} t_{12} \cdots t_{N1} = \text{Trace}[t^N]$$

and as N becomes large, the calculation becomes finding the largest eigenvalue of a 2×2 matrix with

$$t = \begin{pmatrix} e^{-\frac{E(1,1)}{kT}} & e^{-\frac{E(1,-1)}{kT}} \\ e^{-\frac{E(-1,1)}{kT}} & e^{-\frac{E(-1,-1)}{kT}} \end{pmatrix}$$

with 1 indicating a spin up and -1 a spin down. A very elegant approach. Montroll remarked that Ryogo Kubo had developed the same method, but in wartime Japan his contribution was not known to the outside world. In 1974, I met Kubo at a conference in Sitges, Spain. For lunch and dinner meals he would invite a student or two to share his table for conversations and he would order and pay for the wine. Montroll paid for my hotel, but I covered my flight and all other expenses, and wine was not in my budget. Other grad students in Montroll's group would not contemplate using their own money to attend a conference, but I did it again in 1975 for a very fruitful meeting in Antwerpen.

Montroll then went to postdoc with John Kirkwood at Cornell. Kirkwood was considered the dean of chemical physics and trained a large number of students, many of them later worked on the Manhattan Project, including Montroll. Peter Debye was also at Cornell at that time and must have influenced Montroll by his work on lattice vibrations. Montroll used a Green's function approach for lattice vibration problems, and in 1967 he wrote a detailed memorial review of Debye's work in solid-state physics. Back in 1942, Montroll did an exact calculation for a square lattice to discover singularities in the vibrational density of states. These later became known, from a more general work, as Van Hove singularities. The number of singularities depend on where a gradient in wavenumber space goes to zero on a surface with peaks, valleys and saddle nodes. This type of behavior was studied by James Maxwell in 1870 in his paper "On Hills and Dales" about watersheds, which was a general study of the topology of a surface. Maxwell derived a Euler-type formula for (number of peaks)$-$(number of saddle points) $+$ (number of minima) $= 2$. When Montroll much later (I believe at Princeton) on a visit to the library to read Maxwell's collected works, he found that Van Hove had been there and checked out the Maxwell volume. Perhaps, that helped Van Hove to find the Van Hove singularities, or perhaps afterwards he found out about Maxwell's work.

With World War II starting and physicists moving to government work, Montroll found a position as a physics instructor at Princeton University where he met his future wife, Shirley, and they were eventually the parents of ten children, alternating girl, boy five times. Then perhaps through Kirkwood's recommendation, Montroll left Princeton to became head of the math group at the Kellex Corporation with the job of mathematically capturing the behavior of the isotope gaseous diffusion separation process to be built at Oak Ridge. Kellex was a subsidiary of the Kellogg corporation and the X meant it was a secret endeavor. Kellex was founded in 1942 to investigate the feasibility of the gaseous diffusion separation of U-235 from U-238. Kellex was housed in the Woolworth building in lower Manhattan. So, part of the Manhattan project was actually in Manhattan. This is a difficult process to calculate when the system is out of equilibrium. Montroll began with Lord Rayleigh's work on cascade separation of argon from air.

Transient behavior ruled until densities and pressures could reach equilibrium through the many levels of the cascade filtering. At Montroll's 60th birthday I saw a letter from Manson Benedict of MIT (the person who developed the gaseous diffusion separation method for the Oak Ridge plant) praising Montroll's successful work at Kellex.

As non-equilibrium behaviors in density and pressure moved up and down through layers of the cascade this was an example of a random walk on a lattice and the genesis of Montroll's future works and the topic of this book.

From 1939 to 1945 Montroll had major works, the Mayer–Montroll equations, the transfer matrix method for the Ising model, cascade separation processes, exact lattice vibration calculations and works on polymers. In the rest of his life, he fulfilled the promise of his youth with continued advances in physics. And I've only covered some of Montroll's work when he was in his 20's!

In the later 1960s his interests turned to economics and social phenomena, and some of this is captured in his book with Wade Badger "Quantitative Aspects of Social Phenomena" and in his Gibbs Lecture (1987) "On the Dynamics and Evolution of Some Sociotechnical Systems" which was only published after his passing in 1983. Although some at the University of Rochester Physics Department found this interest as orthogonal to a physics department, today these works could be considered the forerunner of what is called econophysics, and Rochester has a Montroll Lectureship. The State of New York created five Einstein Professorships. Rochester was pre-eminent in particle physics, and I was told that the university's choice for the Einstein Chair was in the field of particle physics. But apparently, none of the wives of the selectees were enchanted by the city of Rochester, so instead of getting a third or fourth choice, the department switched fields and Montroll was hired. The President of the university was W. Allen Wallis, a well-respected economist, and Montroll was initiating his work on a physicist's approach to economics, so maybe that played a role in the selection of Montroll? Before getting on to his most cited works on anomalous diffusion in amorphous semiconductors, I note that his course on the Physical Basis of Modern Technology was, after his passing, turned into a book for undergrads by Professor Adrian Melissinos, a high energy colleague in Rochester, so Montroll did plenty to keep the department happy with the physics side of his work. His next book was to be the Physical Basis of Medical Equipment.

Everything about Elliott was positive. It was a delight to know him and to work with him. He was a hands-on mentor always doing his share on joint publications or letting his students publish on their own. When I knocked on his door in 1970, I had no idea how fortuitous and what a life changing experience that would be.

9.2.   A 111100_2th Birthday

In May 1976, I made a surprise 60th birthday party for Montroll in his office. His secretary brought in a rum cake and instead of 60 candles, I used 6, the leftmost 4 lit and 2 remaining unlit creating 60 in base 2 for a savings of 54 candles. John Bardeen was visiting to give a series of lectures on superconductivity, so I invited him. I was expecting Bardeen's talks to be a beautifully laid out theoretical analysis of superconductivity of his Nobel Prize winning work with Leon Cooper and J. Robert Schrieffer. Instead it was a rather dry talk focused on materials that might have a strong excitonic interaction to produce a high temperature superconductor. I met the rest of the Nobel Prize BCS recipients on other occasions. Schrieffer joined my physics division review panel, one year, at the Office of Naval Research. Cooper was well known for his discovery that electrons (through normally repulsive) could pair through lattice vibrations (phonons) at low enough temperatures. They are called Cooper pairs. In 1992, I gave the Michelson Lecture at the US Naval Academy. My topic was chaos and fractals. Some years later Cooper gave the Michelson Lecture, and prior to the lecture I was invited to dinner at the Provost's House along with Cooper and two female midshipman who were identical twins. I never asked, but I think the twins were invited to represent Cooper pairs, or maybe they were the most interested physics majors? Getting back to Montroll's birthday, the 15th Eastern Theoretical Physics Conference was held at the University of Rochester, in November 1976, and immediately following it, was a conference in honor of Montroll's 60th birthday. He gave a lecture showing that he was six handshakes away from Benjamin Franklin. Somewhat recently, it was shown that any two actors could find a set of six movies that connected them through intermediaries. Similar remarks have been made about moving between any two websites through six clicks. But back in 1976 Montroll presented the timeline version of this idea.

During a lunch break, at the Montroll birthday conference, I joined John Ward (the Ward of Ward identities in quantum mechanics). Ward ordered a pitcher of beer with his meal and I thought, OK I'll share the beer with him. Well, he drank the pitcher and ordered a second pitcher which he also drank and spoke about how his work on group theory and gauge symmetries was given less recognition then for those that followed him. Well, he did win a cluster of awards in the 1980s including the APS Heinemann Prize for Mathematical Physics in 1982, but not the Nobel Prize. Later, when I did visit Australia, hearty beer drinking at lunch was not unknown. Montroll did significant works with Ward on finite temperature quantum Green's functions and on the 2D Ising model correlation function using a Pfaffian approach. Montroll considered Ward to be a genius. At Montroll's 60th birthday, his wife was asked at a dinner party of what is it like to be married to a genius. Her reply was, as Elliott is so smart he is better equipped to answer that question. The Ising model correlation paper's third author was Renfrey Potts from Australia and known for the famous Potts model. He spent a sabbatical with Montroll at the University of Maryland,

1955–1956, and tells the story that he went to the stadium to watch the Maryland football team practice as football was a unique American sport. He was immediately accosted by security and accused of being a spy for a competing team. It was obvious where the university set its priorities.

Four years earlier I went to the Eastern Theoretical Physics Conference in Blacksburg at Virginia Tech. In October 1972. I was enjoying the Virginia summer-like weather, while Rochester was quickly approaching winter. Back in those days this was an important annual conference. This one was attended by several Nobel Laureates, all friendly and approachable. We were a group of graduate students driving down together from Rochester, with Montroll footing the bill, and we sat together at the conference dinner. Our table had one empty seat, and Eugene Wigner came over and asked if he could join us. At one point, he went around the table asking what each of us what we were working on. I was in my third semester, pre-qualifying exam, and still taking courses. I mumbled something about random walks and our group, all junior grad students, didn't acquit themselves much better. Wigner looked disappointed not to hear about great problems and advances from our generation. Some years later, I was at a conference at the University of Maryland and the speaker mentioned a Wigner boost. Wigner, who was known for being polite, asked "What is a Wigner boost?". After the explanation, Wigner said he didn't understand and perhaps they could discuss this further after the talk. Not my field, but the speaker I believe was looking at an accelerating reference frame. It should be noted that Wigner had early work, in 1939, on a combination of Lorentz transformations that result in a velocity change (boost) and a rotation, so he should have recognized a Wigner boost better than anyone. He must have had a deeper question for the speaker.

CHAPTER 10

Continuous Time Random Walks

10.1. Time Enters in Probability

Time entered random processes, not as a continuous variable, but discretely as the number of throws of dice or the number of hands played in a card game. Initial probability calculations were focused on combinations and permutations and on independent events. A favorite question is what is the number of people in a group to have a probability over $1/2$ that two people share a birthday. When time is explicitly involved the random process is called a stochastic process. Louis Bachelier introduced a diffusion-type random walk in his 1900 doctoral thesis, "Theorie de la Speculation" to model the Paris stock market changes as a function of time. He included a velocity of change and as we will show later this leads to the telegrapher's equation which is similar, but different from the diffusion equation, as the probability cannot propagate faster than the velocity. The diffusion equation has a probability for being everywhere at short times even through this probability decays rapidly with distance. However, even sticking with the diffusion equation there are mathematically intricate aspects. The path of a Brownian particle has the velocity (tangent to the path) being infinite everywhere. This was eventually handled by Norbert Weiner with his Weiner measure, but at the time when Bachelier was applying for a job Paul Levy found fault with some of Bachelier's mathematics, costing him a position. Nevertheless, Bachelier's work was groundbreaking and preceded the work of Einstein and influenced the works of Kolmogorov. Today it can be considered the birth of econophysics. Einstein in 1905 developed his theory of Brownian motion with velocity ignored and with the diffusion equation derived. Einstein's approach was motivation and genius was for his obtaining Avogadro's number as a factor in the diffusion constant that could be experimentally measured. Langevin followed Einstein's work with a dynamical equation of motion with a white noise term thus introducing the field of stochastic differential equations. We won't follow that ever-expanding field in this essay or discuss correlated "colored" noise.

We will focus on time, not in the dynamical sense as flowing forward at a constant rate, but as a random variable governing the time between events. Waiting time was a topic of queuing theory developed by Erlang, in 1909 in Denmark, to study telephone networks to optimize the time to have a call in a queue connected by an operator. Constant rate processes for adding to a queue and servicing a queue were later generalized to non-Poisson queue properties. Markov invented Markov chains, in 1913, to study dependent events. He took the Puskin novel "Eugene Onegin" and, over a number of pages, counted the number of consonants C and vowels V to find the probabilities of picking a letter at random to be a consonant or a vowel, $p(C) = c$ and $p(V) = v$. If the letters were truly random then the probability of two vowels in a row would be v^2, but that was not what was found, instead $p(VC) > p(VV)$. Markov constructed the probabilities as in the diagram below of pairs VV, CC, VC,

CV to show that finding a V or C next is not independent, but depends on the preceding letter type.

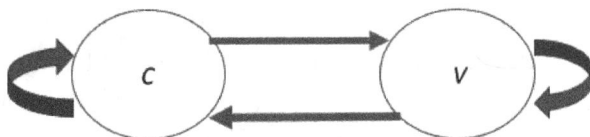

The transitions among these states became known as Markov chains. While Markov used probabilities for the transitions between states, what if there was a time lag for making transitions among the Markovian states. That is precisely the point addressed by Paul Lévy in 1954 at the inaugural International Congress of Mathematicians in his article "Processus Semi-Markoviens." Unlike the early gambling questions to determine odds, Lévy was concerned with theorems about limit distributions. He wrote, "Chaque fois que le systeme etudie arrive a une etat A_n, la duree U_n des un sejour dans cet etat est une variable aleatoire non-negative", my translation, (each time the system under study arrives at a state A_n the duration U_n of its sojourn in that state is a non-negative random variable). With Lévy's paper, waiting times in Markov processes made their debut and the process was dubbed semi-Markovian. Lévy's transitions between states was given in terms of rates instead of probabilities. A fixed rate implies an exponential distribution of waiting times. Fixed rate waiting times were introduced again, independently, in 1955 by W. L. Smith in his article in the Proceedings of the Royal Society "Regenerative Stochastic Processes". Waiting times were to be introduced again in an elegant paper by Elliott Montroll and George Weiss in 1965 for random walks on a lattice.

10.2. A Short Meeting with Elliott Montroll

At my start of graduate school, in September 1970, I knocked on the door of Montroll's office to introduce myself. There was no reply, so for the sake of knocking I knocked louder. Then I heard a voice saying come in. I later learned that I should have entered through the reception area door where his secretary had a desk. Nevertheless, he was gracious, asked me about what math courses I took at Stony Brook and asked me to come back at the end of the second semester. Returning in May, he went to his filing cabinet and took out a half dozen papers. He said, look these over and see if any would interest you. One was the 1965 Montroll–Weiss paper and our meeting was over and I had financial support for the summer.

I read all the papers that Montroll selected and the Montroll–Weiss one was my favorite from the bunch. That paper was a masterful calculation of evermore complicated quantities, such as first passage time moments, the number of points visited r times in an n-step walk, and exact probabilities for return to the origin for a variety of 3D lattices (while Feller's book still used approximate values, despite Montroll saying that he communicated the exact results to Feller). It is only towards the end, in Sec. V, that the continuous time random walk (CTRW) is introduced. While the formalism was elegant, the main result was that n, the number of jumps, was replaced by $t/\langle t \rangle$ with $\langle t \rangle$ being the mean waiting time between jumps and t is the time of the ongoing process. The 1965 paper attracted no attention until the 1973 work of Scher and Lax (more about that later, on ac impurity conduction in an amorphous semiconductor), that considered a hopping process where the mean waiting time for a jump was longer than the time of the experiment. That brought in new physics, "fractal time" and many avenues of exploration, so many new works that in fact that the 50th anniversary of the Montroll–Weiss paper was celebrated by the volume with twenty articles in, "The Continuous-Time Random Walk, Still Trendy: Fifty Year History, Current State, and Outlook" in the *European Physics Journal B.*

After our brief meeting, Montroll left for the summer of 1971. He had the JASON DOD Advisory group summer meeting in La Jolla and afterwards family time at his summer cottage at Woods Hole in Cape Cod. I was paid a salary, $300 a month (rent was $75 a month at the Graduate Living Center for a shared two-bedroom apartment on the landing flight path to the nearby airport), and I was free to follow my interests. Most of my fellow grad students worked long hours in labs under faculty supervision. I spent the summer enjoying careful reading of Feller's Vol.I and most of Vol.II, the Wax reprint volume of the classic papers on noise and stochastic processes, by Chandrasekhar, Uhlenbeck and Ornstein, Wang and Uhlenbeck, Rice, Kac, and Doob. Now onto the what exactly is the CTRW.

10.3. The Montroll–Weiss Continuous Time Random Walk

The CTRW begins with $Q(l, t)$ the probability density to reach site l exactly at time t, and $\psi(t)$ the probability per unit time that the walker waits a time t between jumps. The equation for $Q(l, t)$ is given by,

$$Q(l, t) = \sum_{l'=-\infty}^{\infty} \int_0^t Q(l - l', t - \tau)p(l')\psi(\tau)dt + \delta_{l,0}\delta(t)$$

(10.1)

Fourier transforming to k-space and Laplace transforming to s-space yield

$$Q(k, s) = \frac{1}{1 - \psi(s)p(k)}$$

(10.2)

which equates to the Greens function $G(k; z)$ where the dummy variable z has been replaced by $\psi(s)$. The probability $P(l, t)$ occurs when the walker reaches site l at time $t - \tau$ and then does not jump for at least a time τ. This gives the famous Montroll–Weiss formula

$$P(l, t) = \int_0^t Q(l, t - \tau)\Phi(\tau)d\tau \qquad (10.3)$$

with $\Phi(\tau) = \int_\tau^\infty \psi(t')dt'$ the probability that the next jump takes place after time τ.

In Fourier–Laplace space we have the most used form for the Montroll–Weiss CTRW

$$P(k, s) = \frac{1}{1 - \psi(s)p(k)} \frac{1 - \psi(s)}{s} \qquad (10.4)$$

The author at home in Rockville, MD in March 1983 between George Weiss and Elliott Montroll (with jacket and tie).

10.4. A Master Equation for the CTRW

Montroll's grants supported ten graduate students (I became the tenth in 1971) and ten postdocs, plus he helped support two faculty members, one visiting from Brazil and one a former grad student. He had a whole wing on the second floor of the Bausch and Lomb Physics building on the quadrangle. He taped a piece of cardboard on the window pane of the entrance door to the wing with the handwritten words "Institute for Fundamental Studies" which is what he called his group. The cardboard looked like it was taken from the packaging of a shirt. He took the cardboard off the window when he moved from the University of Rochester.

As he was paid by New York State through an Einstein Professorship, Montroll was free from department responsibilities such as teaching and committee assignments. He did teach a course called "Quantitative Aspects of Social Phenomena" that was a forerunner of what today would be called econophysics or complexity theory. The lectures were published as a book with his grad student Wade Badger. It should have been a classic, but is not well known. He also taught courses on modern technology (his notes posthumously turned into a book by Adrian Melissinos "The Principles of Modern Technology"), and courses on statistical physics and nonlinear dynamics. It disturbed some physics faculty that Montroll was free from university commitments and not focused on physics, but on social phenomena. The University of Rochester was perhaps best known at the time for high-energy particle physics and participating in the big experiments at national labs. Rochester had spectacular sunsets (perhaps due to the chemicals used at the Kodak plant), but with its long and harsh winters (but great for cross country skiing) it had to compete for faculty with other perhaps more attractive cities. The city was also anchored by Xerox Research Labs in nearby Webster and Bausch and Lomb for lenses and optical equipment. The Physics Department, through its chairman Robert Marshak, hosted the famous Rochester conferences in particle physics and the faculty was outstanding and tilted towards this field. It was said the Einstein Chair was first offered went to top flight particle physicists, but their spouses rejected the offer and the city. It was now 1966 and a good full professor salary was around $30,000. The Einstein salary was $50,000 plus $50,000 for expenses like travel and secretarial help. Finally, the Rochester Dean supposedly said I don't want your fourth choice, let's switch fields and the offer was made and accepted by Montroll who had a stellar reputation in physics going back to the 1940s. With ten children, even though the salary must have been attractive, he spent much of his time consulting to further support his large family. He typically only came in one day a week as he was out consulting or working from home. I never remember him at a faculty meeting and only at talks if he knew the speaker. All of this led some professors to disparage Montroll and advise new grads students to steer clear of Montroll as he wasn't focused on physics. Fast forward to the present where Montroll's most cited work is on the physics of fractal time random walks from his 1970s work at Rochester that arose from his consulting at Xerox and on stretched exponential relaxation in glassy materials that arose from his consulting at General Electric. Today the University of Rochester Physics Department has a Montroll Memorial Lecture. My PhD arose from the work at Xerox and I inherited Montroll's consulting position at General Electric.

In 1971 I was in my second year of graduate school, taking courses and disappointed in my interactions with Montroll. There were none as Montroll just wasn't around and he had all the other more senior grad students and postdocs. After reading the Feynman Lectures as an undergrad I felt that I was a physicist and wanted to be treated as such. Anyway, the qualifying exam was coming up in June 1972 and that's what I needed to focus on. After a faculty meeting a new schedule was announced, the exam would be given in March 1972, not June. Most grad students in my class decided not to take the fast approaching March exam. I decided to take the chance with the reward that if I passed I would have the summer to do research and not spend it studying. I believe I can understand and learn anything in physics given the time to study it, take it apart, and put it back together. I can also draw a blank if asked to do that during a test. Some problems on the qualifying exam I could do and some I could kick myself for not figuring out to set them up. I passed by the skin of my teeth and the summer was mine. Montroll was again away for the summer.

In 1973, I stopped by Montroll's office to chat. He posed a problem. He had shown how to write the continuous time random walk as a master equation as long as the mean waiting time between jumps was finite. He was considering the case when the mean wait time was infinity. His approach failed when that mean time is infinite and said that would be a good problem to fix. Nitant Kenkre, a postdoc, liked to listen in to discussions in Montroll's office and he bounded in to join our talk as his field was generalized master equations. We all went on our way and within a day we all solved the problem where the key was to introduce a memory function and we shared in the publication. Start with this master equation,

$$\frac{\partial P(l,t)}{\partial t} = -\int_0^t P(l, t-\tau)\varphi(\tau)d\tau$$
$$+ \sum_{l'} \int_0^t P(l-l', t-\tau)p(l')\varphi(\tau)d\tau \quad (10.5)$$

where $\varphi(t)$ has units of $1/t^2$ and we want to related it to the probability waiting time density $\psi(t)$. By Laplace transforming (over time) and Fourier (over space) in Fourier–Laplace space, the master equation is in the form,

$$sP(k,s) - P(k,t=0) = \varphi(s)p(k)P(k,s) - \varphi(s)p(k) \quad (10.6)$$

Solving for the transformed probability gives

$$P(k,s) = \frac{1}{s + \varphi(s) - \varphi(s)p(k)} \quad (10.7)$$

If we set

$$\varphi(s) = \frac{s\psi(s)}{1 - \psi(s)} \quad (10.8)$$

then

$$\begin{aligned}
P(k,s) &= \frac{1}{s + \frac{s\psi(s)}{1-\psi(s)} - \frac{s\psi(s)}{1-\psi(s)}p(k)} \\
&= \frac{1 - \psi(s)}{s(1 - \psi(s)) + s\psi(s)(1 - p(k))} \\
&= \frac{1}{1 - \psi(s)p(k)} \frac{1 - \psi(s)}{s}
\end{aligned}$$

which is the CTRW result.

Now $\varphi(t) = a\delta(t)$ transforms to $\psi(t) = a\exp(-at)$. When $\varphi(t)$ takes the simple exponential form, $\varphi(t) = a\exp(-bt)$ which transforms to $\varphi(s) = \frac{a}{b+s}$, this generates a somewhat more complex,

$$\begin{aligned}
\psi(s) &= \frac{\frac{a}{(s+b)}}{s + \frac{a}{(s+b)}} = \frac{a}{s(s+b) + a} \\
&= \frac{a}{\left(s + \frac{b}{2} + \frac{1}{2}\sqrt{b^2 - 4a}\right)\left(s + \frac{b}{2} - \frac{1}{2}\sqrt{b^2 - 4a}\right)}
\end{aligned}$$

We can write $\psi(s)$ as a sum of two partial fractions, each transforming back as an exponential to arrive at,

$$\psi(t) = \frac{2a}{\sqrt{(b^2 - 4a)}} \exp(-bt/2) \sinh\left(\frac{1}{2}t\sqrt{(b^2 - 4a)}\right)$$

Note if $b^2 - 4a < 0$, then the sinh becomes an i sin and the i cancels from the i coming from the square root in the denominator to keep $\psi(t)$ real. So, a simple $\varphi(t)$ transforms into a more complicated $\psi(t)$.

I thought that just rewriting the CTRW as a generalized master equation accomplished nothing new and wasn't worthy of publication. I was wrong, and the citations to this work continues to grow.

10.5. Moments of the CTRW

Previously, we showed how to calculate moments in terms of the number of steps taken. Now we repeat that for the CTRW in terms of the time taken and use the transforms $x \to k$ and $t \to s$ with $\mathcal{L}^{-1}$ being the inverse Laplace transform

$$\langle x(t) \rangle = \sum_x xP(x, t) = -i\frac{\partial}{\partial k|k = 0}\sum_x \exp(ikx)P(x, t)$$

$$= -i\frac{\partial}{\partial k|k = 0}P(k, t)$$

$$= -i\mathcal{L}^{-1}\frac{\partial}{\partial k|k = 0}P(k, s)$$

$$= -i\mathcal{L}^{-1}\frac{\partial}{\partial k|k = 0}\frac{1 - \psi(s)}{s}\frac{1}{1 - \psi(s)p(k)}$$

$$= \bar{x}\mathcal{L}^{-1}\frac{1}{s}\frac{1}{1 - \psi(s)} \tag{10.11}$$

where we have used

$$\frac{\partial p(k)}{\partial k} = \frac{\partial}{\partial k} \sum_x \exp(ikx) p(x)$$

$$= i \sum_x \exp(ikx) x p(x) \qquad (10.12)$$

and evaluating at $k = 0$ gives $i\bar{x}$ where $\bar{x}$ is the average single jump magnitude.

We'll look at the long-time asymptotic limit, so small s in Laplace space for,

$$\psi(s) = \int_{t=0}^{\infty} \exp(-st) \psi(t) dt \qquad (10.13)$$

goes as, when the moments of $\psi(t)$ are finite, $\psi(s) \sim 1 - s\langle t \rangle + O(s^2)$, therefore

$$\langle x(t) \rangle \sim \bar{x} \mathcal{L}^{-1} \frac{1}{s^2 \langle t \rangle} = \bar{x} \frac{t}{\langle t \rangle} \qquad (10.14)$$

Nothing surprising, just the number of steps is replaced by $\frac{t}{\langle t \rangle}$ I won't go through all the steps for higher moments, but will just give the result,

$$\langle x^2(t) \rangle = \sum_x x^2 P(x,t) = -\frac{\partial^2}{\partial^2 k} \Big|_{k=0} \sum_x \exp(ikx) P(x,t)$$

$$= \bar{x}^2 \mathcal{L}^{-1} \left\{ \frac{\psi(s)}{s(1 - \psi(s))} \right\} + 2\bar{x} L^{-1} \left\{ \frac{\psi(s)^2}{s(1 - \psi(s))^2} \right\}$$

$$(10.15)$$

For the case when the first and second moments of $\psi(t)$ are finite, we have for the variance,

$$\sigma^2(t) = \langle x^2(t) \rangle - \langle x(t) \rangle^2 = \left\{ \frac{\bar{x}^2}{\langle t \rangle} + \left(2\frac{\bar{x}^2}{\langle t \rangle} \right) \left(\frac{1}{2} \frac{\langle t^2 \rangle}{\langle t \rangle^2} - 1 \right) \right\} t$$

$$(10.16)$$

10.6. CTRW First Passage Times and Sites Visited

We can revisit other quantities, now in the CTRW context, such the number of sites distinct visited $S(t)$ by time t instead of S_n the distinct sites visited after n steps. This will be an important variable for motion in a disordered medium when the mean waiting time between jumps is infinite in what we call fractal time:

$$S(t) = \sum_r \int_0^t F(r, \tau) d\tau \qquad (10.17)$$

We calculate,

$$F(r, t) = \sum_{n=0}^{\infty} F_n(r) \psi_n(t) \qquad (10.18)$$

where $\psi_n(t) = \int_0^t \psi(\tau) \psi_{n-1}(t - \tau) d\tau$ is the probability density for the nth jump to be at time t. Note that,

$$\psi_n(s) = [\psi(s)]^n$$

making the Laplace transform for $F(r, s)$ into a generating function,

$$F(r, s) = \sum_{n=0}^{\infty} F_n(r)[\psi(s)]^n \qquad (10.19)$$

We proceed by taking the Laplace transform of $S(t)$,

$$\mathcal{L}S(t) = \frac{1}{s} \sum_r F(r, s)$$

$$= \frac{1}{s} \frac{1}{G(0, \psi(s))} \left(\sum_r G(r, \psi(s)) - 1 \right) \qquad (10.20)$$

where we have used Eq. (1.26) $F(l, z) = \frac{G(l,z) - \delta_{l,0}}{G(l=0,z)}$ with z replaced by $\psi(s)$ and using $\sum_r G(r, \psi(s)) = \sum_r P_n(r)[\psi(s)]^n = \frac{1}{1 - \psi(s)}$

yields,

$$S(t) = \mathcal{L}^{-1}\left[\frac{1}{s}\frac{\psi(s)}{(1-\psi(s))G(0,\psi(s))}\right] \qquad (10.21)$$

so basically, all we have worked through was changing from the number of steps to time by changing z to $\psi(s)$ in the Green's function. When $\langle t \rangle$ is finite, for long times (i.e. small s), $1 - \psi(s) \sim s\langle t \rangle$ and we recover the mean number of steps result with n replaced by $t/\langle t \rangle$.

10.7.   Entering an Ongoing CTRW Stochastic Process

Mel Lax, a master of stochastic processes, classical and quantum, liked to pose the type of question of entering a room with a light bulb turned on that's rated to last, on the average, 1000 hours. You don't know how long the light has been on and now want to guess when it will burn out. Let's study this in terms of a random walk. This question arose in the Scher–Montroll model of photocurrents in amorphous materials with a distribution of trap depths, such that the average trap time was infinite. In the experiment, a flash of light creates exciton pairs whose electrons and holes move in opposite directions. Charges that get trapped in shallow traps leave and tend to then get trapped in deeper traps and so on. As times goes on the charges find themselves in the deepest traps for times longer than the experiment. So if one entered this as an ongoing process that could have started long ago the math would say the average charge is trapped and no movement is occurring. This type of result was contrasted with the CTRW result. The experiment in question started with a flash of light that set a time $t = 0$ so one calculated from the start of a random process. Anyway, we'll derive how to treat the first step of a random walk when you enter an ongoing random process.

Assume the random walker is on a lattice site and you do not know how long it has been sitting there, but you do know $\psi(t)$ the probability density for the waiting time between jumps.

Start with the equation for the mean squared displacement for a random walk with no bias,

$$\langle x^2(t)\rangle = \bar{x^2}\mathcal{L}^{-1}\left\{\frac{\psi(s)}{s(1-\psi(s))}\right\}$$

$$= \bar{x^2}\mathcal{L}^{-1}\frac{\psi(s)}{s}[1+\psi(s)+\psi^2(s)+\psi^3(s)+\cdots]$$

As an example, choose $\psi(t) = \lambda^2 t \exp(-\lambda t)$ whose Laplace transform is $\psi(s) = \lambda^2/(s+\lambda)^2$ and this gives,

$$\langle x^2(t)\rangle = \frac{1}{2}\bar{x^2}\left[\lambda t + \frac{1}{2}(1-\exp(-2\lambda t))\right]$$

which has the asymptotic long time limit is $\frac{1}{2}\bar{x^2}\lambda t$. Can we choose a first step with a different waiting time probability to achieve the asymptotic result for all times. Note that $\langle t\rangle = \frac{2}{\lambda}$ What if we observe the walker and do not know how long it has been at the origin. We need to take into account the probability that it has there all the way back to t equals minus infinity. We need to treat the first step differently than the succeeding ones. Let us call the first step probability density $h(t)$. Then,

$$\langle x^2(t)\rangle = \frac{1}{s}\bar{x^2}\mathcal{L}^{-1}[h(s)+h(s)\psi(s)+h(s)\psi^2(s)+\cdots]$$

$$= \bar{x^2}L^{-1}\frac{h(s)}{s}\frac{1}{(1-\psi(s))}$$

Choosing the waiting time probability density for the first jump to be,

$$h(s) = \frac{1-\psi(s)}{s\langle t\rangle} \tag{10.22}$$

gives

$$\langle x^2(t)\rangle = \bar{x}^2 \mathcal{L}^{-1}\frac{1}{s^2\langle t\rangle} = \bar{x}^2\frac{t}{\langle t\rangle} \tag{10.23}$$

the asymptotic limit as long as $\langle t\rangle$ is finite. Not knowing the starting time brings one to the asymptotic result for all times. I think Feller might have been the first one to address and derive the $h(t)$ expression. Note for a Poisson process $\psi(t) = \lambda\exp(-\lambda t)$ then $\psi(t) = h(t)$ You can see the equation shows that $h(s) = 0$ when $\langle t\rangle = \infty$.

For $s \to 0$,

$$h(s) = \int_0^\infty \exp(-st)h(t)dt \sim 1 - s\langle T\rangle \tag{10.24}$$

and using Eq. (10.24) for $h(s)$

$$h(s) \sim \left(1 - \left(1 - s\langle t\rangle + s^2\frac{1}{2}\langle t^2\rangle\right)\right)\Big/s\langle t\rangle \sim 1 - s\langle t^2\rangle/2\langle t\rangle$$

For Poisson, $\langle t^2\rangle = 2\langle t\rangle^2$. In general when moments exist, the mean waiting time $\langle T\rangle$ for the first step for observing an ongoing random process is $\langle T\rangle = \langle t^2\rangle/2\langle t\rangle$.

In a footnote of volume II of his classic book, Feller remarked that outside of queuing theory it's hard to find examples besides a bus running in a circular loop without a schedule that call for non-Poisson renewals. However, even for a random walk on a lattice with Poisson distributed waiting times between jumps, questions can be posed that lead to distributions with long heavy tails, such as first passage time problems. So, distributions with long-time tails were well known for diffusion problems when the right question is posed. However, a direct long-time tail probability for waiting between jumps was only introduced in 1973 by Scher and Lax to model ac conductivity hopping processes in an amorphous semiconductor and then applied by Scher and Montroll to explain dispersive transport in xerographic films. Mathematically, a waiting time distribution can have an infinite mean, but in physical models a cut-off will be always be applicable and the question that becomes is the experimental measurement time larger or smaller than the cut-off time. We will present the role of the CTRW in later chapters on the Scher–Lax, Scher–Montroll papers, and on the glass transition.

Below are some references to a few historical papers.

A.K. Erlang, "The Theory of Probabilities and Telephone Conversations", *Nyt Tidsskrift for Mathematik B* 1909 vol 20, 33–39

A.A. Markov, *Bull. Imp. Acad. Sci. St. Petersburg* **7**, 153 (1913)

P. Lévy, *Proceedings from the International Congress of Mathematicians* **3**, 416 (1954)

W.L. Smith, *Proc. Roy. Soc. Ser. A* **232**, 6 (1955)

W. Feller, *An Introduction to Probability Theory and Applications*, Vol. II (1966) Footnote on page 186.

H. Scher, M. Lax, *Phys. Rev. B* **7**, 4491 (1973)

H. Scher, E.W. Montroll, *Phys. Rev. B* **12**, 2455 (1975)

CHAPTER 11

Conferences

11.1. Random Walks Then Fractals Then More

In 1975, at a conference in Antwerpen, I met Yossi Klafter who was a graduate student at Tel-Aviv University. He latter post-doc'ed at MIT and when he was set to return home, his MIT advisor, Robert Silbey got a call. It was from one of his former postdoc's now working for Exxon, in Linden, NJ. She said that Exxon was building a basic research center and was looking for suggestions on whom to hire. Yossi moved to Exxon and invited me to give a talk. Not sure if they were refining oil or just storing large quantities, but the air had an unmistakable smell of oil and I think sulfur. My talk was about fractals in condensed matter, then a somewhat new topic, and it went over very well. The next thing I knew, was that I was invited back for a job interview, although I was not interested. By this time, Exxon research had moved to a brand-new facility in rural Annandale, NJ. The story was the lab was built on farm land and Exxon was paid by the Agricultural Department for not planting crops. I gave another talk and went to dinner with Yossi, Roger Cohen the lab director, and Morrell Cohen the Chief Scientist. As Morrell was a wine connoisseur, Roger asked him to choose the wine. He picked the most expensive one from the list. Roger shocked by the price, whispered to Morrell if he wanted to reconsider the order. Morrell said, you're right and called back the wine steward and said, I ignored how many people we are, so make it two bottles. Money flowed freely in those days. Exxon was aiming to hire 200 scientists and engineers and were calling themselves the Bell Labs of energy. A fair number of people began to work on fractal aspects of materials, and on oil recovery. This was 1982, and to skip ahead, life at the lab was great with young scientists, visitors, seminars and conference trips with travel reimbursements for high living. Bob Zwanzig had warned me that there is a resonance when you get hired by industry, it fits both of you. But industry plans and managers can change quickly and the resonance disappears. This was true at Annandale, as in 1986 Exxon closed down most of the Annandale facility. At my last visit there, most of the space was now occupied by pharmaceutical companies. Apparently, Exxon shifted their effort to the more applied Exxon lab in Houston.

But back to the Exxon dinner. As a young research scientist, I was not an invited speaker at conferences, so I came up with the idea to hold my own conference about all aspects of random walks. Where to get funding? I asked Roger Cohen for $10,000 to fund a meeting on random walks with a focus on fractals. He gave an unofficial nod. Through contact with Bob Rubin at the National Bureau of Standards (latter called NIST) in Gaithersburg, MD I arranged to hold the meeting. (Note Bob's wife Vera Rubin was famous for the dark matter hypothesis based on galaxy rotation observations. Vera was our first Montroll Memorial Lecturer.) I got George Weiss from NIH to join me in organizing the meeting. George was an early Montroll grad student and the co-authors of the famous continuous time random walk paper. George once remarked that his Army duty wasn't so bad. On the garbage detail, sometimes the sergeant would let him sit inside the truck if it was raining. Anyway, the Exxon contribution only came to $1000 and George managed to come up with the rest from NIH. Harvey Scher contributed some funds from Xerox. We were off and running and around 200 people attended. I invited Benoit Mandelbrot as the after-dinner speaker.

From not being an invited speaker, after this random walk meeting, I have ended up, so far, giving around 400 invited talks, so my plan worked. The Washington area had a rich set of research labs, including NIH, NIST, NRL, NSWC White Oak and Carderock, University of Maryland, and Johns Hopkins APL, along with the major government funding agencies, making the Washington area a wonderful research environment and location for a conference. I embarked on organizing my next meeting with support from the Office of Naval Research. The topic was Fractals. By a curious set of circumstances, I ended working for ONR before the start of the meeting. So, in the Fall of 1983, with Benoit Mandelbrot, we jointly held the first US conference on Fractals, again with NBS as the venue. I did the organizing and asked Benoit what he was going to do, his answer was "invite the press". He had his own successful plan about recognition and public relations. After that large NBS fractals meeting, there commenced a growing list of fractal meetings. The next one was held a few months later at the famous French conference center in Les Houches. I was listed as an organizer, but did nothing in the way of organizing. I didn't even know how to get to the conference site. It began on a Monday and after landing in Geneva and taking a bus, I arrived at Les Houches on a Sunday night along with some other participants. I was expecting some information upon arrival of exactly where was the meeting, but it was snowing and there was nothing in sight. Everything was closed in the town. Happily, a cab was found and it turned out the conference center was part way up on the mountain road with Chamonix nearby. So, all was well in this first trip to Les Houches, and I was an old hand for later meetings there. While I tried my hand on the green slope, I discovered that Fereydoon Family, then at Emory University, was an expert black diamond skier. He was a leader in the new field of fractals and did beautiful work on real space renormalization, fractal surface growth, and the shape of snowflakes.

Another conference in Szeged, Hungary was the opposite in that perfect travel instructions were provided. It was organized by Laszlo Kish as the *Unsolved Problems of Noise Conference*. Laszlo insisted on double blind refereeing to determine the speaker list. His meeting continues till today. My flight from the US was delayed, so instead of landing in Frankfurt in the morning and catching a flight to Budapest, I arrived in the late afternoon instead. The later flight to Budapest got me in at night. Next, I needed to catch a train and the instructions were clear about which train station to use, but the airport only had shared shuttles instead of taxis. My shuttle dropped off the other passengers first at a variety of hotels in downtown Budapest, and finally myself at the train station. I didn't see my train on the announcement board, but a station employee directed me to a lower level, and I got on a line to purchase a ticket. Finally, running to the train I hopped on board just as it began to move. I believe they were actually waiting for me to board. Szeged was the last stop and the train station was closing with lights shutting off. A station employee helped me find a taxi and I got to the conference hotel after midnight. It all worked thanks to the help of railway staff. After traveling alone for most of the day and night, I entered the hotel and immediately saw friends, a great feeling.

There were so many interesting conferences. The one at Geilo, Norway asked for your ski boot size as you could ski from the hotel to the slopes. In Lanzhou, China, we afterwards toured the Jiayuguan Silk Road Fortress and the DunHuang Buddhist caves in the northwest. In Kyoto, in 1991, I took the hotel elevator down to meet my friend Takashi Odagaki when it started to bang into the walls on the way down and then stopped. I thought it strange that this fine hotel had a poorly working elevator. When the elevator soon started up again and I got to the lobby my friend was not yet there, and didn't come, only arriving over a good bit later. The first thing he said was, where were you when the earthquake hit. I had no idea. He was in the subway and that was a more worrying affair than mine. Takashi did beautiful work with Mel Lax on dynamical properties of disordered materials, more about that in Chapter 15. With many outstanding works, his latest is modeling the effects of quarantining on the covid-19 virus spread.

With Takashi Odagaki and Harvey Scher in Kyoto.

After one conference, I got a letter from the Swedish Academy of Sciences. My guess was an invitation to nominate for the Nobel Prize, instead, surprise, it was a bill from Physica Scripta for a paper that I never submitted. I turned out that a conference was publishing a proceeding to which I submitted a paper. The two main famous speakers did not submit papers and the organizer lost interest and shipped the submitted papers off to Physica Scripta where they were published as part of an issue along with non-conference papers, and hence the surprising bill. As a grad student at the University of Rochester, I volunteered to post the announcements, at the relevant department bulletin boards, for courses of the Ettore Majorana Centre for Scientific Culture in Erice, Sicily. In December 1979, I attended one of those courses, as I was leaving Georgia Tech and was on my way to spend time at Tel-Aviv University before moving to the University of Maryland. I had no financial support so everything was out of pocket. The organizer Dr. B. di Bartolo, from Boston College, waived my registration fee by considering me part of the Centre's family as I was their U. of Rochester representative. A great feeling.

11.2. Cargese and Searching Lévy Style

The physics world has a number of "watering holes". In addition to Les Houches, the French have Cargese on the island of Corsica. The first of our Cargese fractal meetings was organized by Gene Stanley from Boston University and Nicole Ostrowski from the University of Nice. Gene is well known for being at the center of an international network of students, postdocs, and colleagues working together on topics in statistical physics. He is a man of many talents. Once in Rio de Janeiro, walking at night we came upon a dance troupe practicing in the street. Gene joined them, was accepted, and didn't miss a step. Nicole arranged everything and made sure the meeting ran smoothly, at that and then at later Cargese meetings. Cargese is a wonderful venue with lunches on long benches overlooking the Mediterranean, a limited number of talks and time to work and discuss physics. A very friendly atmosphere.

With George Zaslavsky on my right and Fereydoon Family in Cargese, Corsica. George introduced the Zaslavsky map for a kicked electron in a magnetic field. The phase space produced a stochastic web with trajectories with fractal random walk like properties. Fereydoon was a leader in the fractal community with works on real space renormalization and on fractal growth.

When I was a research scientist at Georgia Tech in the late 70s, I had my first thoughts about search when after a colloquium, talking with Philip Morse from MIT about his operations research work during WWII. He described the air search pattern off the west African coast looking for submarines. It was a figure eight. At a later meeting in Minneapolis, Barry Ninham, a famous Australian physicist and an early Montroll student, remarked to me that he saw something about ants performing random walks that might be fractal like. So, when I needed to write something for the Cargese proceedings "On Growth and Form", I had to come up with something new. I contributed the work "Lévy Flights versus Levy Walks" after watching flying insects having seemingly random Lévy-like flights combined upon landing with what seemed like Brownian random walks. Sitting on the beach I worked out the asymptotics of Brownian versus Lévy search for scarce targets showing the benefit of scale invariant motions. My contribution was together with Yossi Klafter. Our example was rather short and simple, but it spurred an industry of ever increasing sophistication of search problems. Here is the analysis for Brownian and Lévy search.

Start with traps randomly placed in d dimensions. The probability that a region of radius R and volume $V \propto R^d$ is devoid of traps, scales as $\exp(-V)$. The probability that a random walker, performing a Brownian-type motion starting in the center of that volume, is still within volume V at time t, scales as $\exp(-t/R^2)$. Averaging over a Poisson distribution of volumes, the lifetime distribution for not being trapped is,

$$\phi(t) \sim \int_0^\infty \exp(-V)\exp(-t/R^2)dR$$

Using the method of steepest descent, first setting

$$\frac{d}{dR}\left[R^d + \frac{t}{R^2}\right] = 0$$

yielding $t \sim R^{d+2}$, and $\frac{t}{R^2} = \frac{t}{t^{\frac{2}{(d+2)}}} = t^{d/(d+2)}$ which has the same scaling as R^d, yielding,

$$\phi(t) \sim \exp(-t^{d/(d+2)})$$

a result of Balagorov and Vaks (*JETP* 38, 968 (1974)) of $\exp(-t^{3/5})$ in 3D.

For our Lévy walk where $\langle R^2(t)\rangle \sim t^2$ so the appropriate term is $\exp(-t/R)$, a similar analysis gives,

$$\phi_{L\acute{e}vy}(t) \sim \exp(-t^{d/(d+1)}) = \exp(-t^{3/4}) \text{ in 3D}$$

a faster decay in 3D over time to encounter a trap (or find a target) than the Brownian search. This is a better scheme if the trap is a food source. It seems this little calculation started a field of search finding optimization for scale invariant search patterns for scarce targets. Many variants have been explored under Lévy processes, such as, using alternating search patterns of Brownian and Lévy, search with a time limit, heterogeneous search regions, search with a return to a home base, swarms of searchers, dense or scarce targets, renewable targets, and moving targets and fleeing targets.

With Yossi Klafter and George Zaslavsky at a chaos conference organized by George in Carry-le-Rouet on the coast near Marseilles.

11.3.   The Mandelbrot 65th and 70th Birthdays

Here, I'll just recount two more meetings, the Mandelbrot 65th birthday and 70th birthday. For the 65th we stayed (off season) at the upscale Mas d'Artigny hotel in St. Paul de Vence in the hills north of Nice, France. The lobby had a grotto with the pool half outside and half inside the hotel. Individual cottages had their own pool. The conference dinner was highlighted by 65-year old cognac. The meeting was a celebration of fractals and the keynote speakers were Michael Fisher and Sam Edwards. Neither speaker was known for work on fractals either before or after the meeting, but the rest of us were eager contributors to the math and applications of fractals. The proceedings are in *Fractals in Physics: Essays in Honor of Benoit B. Mandelbrot,* eds. A. Aharony and J. Feders (North-Holland 1989). Mandelbrot was absolutely brilliant and sharp and could tell in a talk what was novel, what was known and what was suspicious. He would sometimes interrupt speakers to remind them of his work implying that he should be cited. This put off some physicists who did not see a need to reference him in their work. By analogy, if a person invented the game of chess would that person get all the credit for the strategies in great games of chess. After all, all the games follow from the rules. Mathematicians and physicists have different views on how to advance and give credit in their fields.

For Mandelbrot's 70th birthday the meeting was in Curacao, arranged by Carl Evertz, who had worked with Mandelbrot and had family connections in Curacao. The plane from Miami was mostly occupied by our conference participants and we were met upon landing with a reception and with an invitation by the Gouveneur to a banquet. At the first lunch, I saw a rather overweight man sitting by himself and he did not look like he

On the way to Curacao. Benoit and Aliette Mandelbrot in front and the author two rows behind.

belonged. So, I sat next to him and started a conversation. He began to talk at two or three times the natural conversational speed. It turned out he was Carleton Gajdusek, the 1976 Nobel Prize winner for Medicine and a researcher at NIH. In the late 1950s, he had studied Kuru disease in the Fore tribe in New Guinea, a prion protein misfolding disease of the brain transmitted through, as an honor, eating the brain of the deceased. He talked to me of how quickly the children in the tribe caught on to reading and the use of a typewriter. He brought back artifacts from the tribe and had trouble finding a museum to accept them. They had all wanted funds for the upkeep. Many of Gajdusek's collection did end up in the Peabody Museum of Salem, Massachusetts that I toured with Gene Stanley and his wife on one of my many trips to Gene's group at Boston University. Gajdusek's talk was a slide show of times at Cal Tech where he overlapped with Mandelbrot, but he couldn't find any pictures of Mandelbrot. Two years later, following our meeting, Gajdusek was arrested and sent to prison. He brought boys from New Guinea to obtain a western education and be trained as lab technicians to upgrade medical care back in their homeland. Complaints arose from some of the boys of molestation resulting in the prison term. Wikipedia has more information. Later, Mandelbrot in one of our far-ranging discussions dropped the remark that Gajdusek said that he was finally able to lose weight and was feeling better than ever being in prison. On release, he moved to Europe. The final twist is that when Montroll moved from his large house in Chevy Chase, Maryland to the University of Rochester, it was Gajdusek (with all of his boys) who rented and lived in Montroll's house.

CHAPTER 12

Coupled Space-Time Memory Random Walks

12.1. Coupling Jump Length and Time

While the diffusion equation appears at first sight to be a proper description of the diffusion process, at closer inspection troubles appear. At an infinitesimal early time the probability to be at a faraway site is non-zero. If studying pollution and 10^{23} mercury atoms are dumped in the ocean, you do not want to overestimate that some have traveled almost immediately great distances. The remedy is to put a velocity into the formulation of a new equation, and we do that with a random walk with a space and time coupled memory. We will even try to tame the Lévy flight whose single jump mean squared distance is infinity. But first, we'll go from the diffusion equation to the telegrapher's equation. Our approach follows that of George Weiss in his book *Aspects and Applications of the Random Walk*.

We start with,

$$\Psi(x, t) = p(x|t)\psi(t) \tag{12.1}$$

where $p(x|t)$ is the conditional probability of a jump x given that the jump took a time t. We stick to 1D for simplicity, but x can be a vector in any number of dimensions. Examine the case of all jumps having the same velocity V, so

$$p(x|t) = \delta(x - Vt) \tag{12.2}$$

and

$$\psi(t) = \lambda \exp(-\lambda t) \tag{12.3}$$

Fourier transforming gives

$$\Psi(k, t) = \cos(kVt)\lambda \exp(-\lambda t) \tag{12.4}$$

and with Laplace transforming, we arrive at,

$$\Psi(k, s) = \frac{\lambda(s + \lambda)}{(s + \lambda)^2 + (kV)^2} \tag{12.5}$$

Using the CTRW formula

$$P(k, s) = \frac{1}{1 - \Psi(k, s)} \frac{1 - \psi(s)}{s}$$

yields,

$$P(k, s) = \frac{(s + \lambda)^2 + (kV)^2}{s^2 + s\lambda + (kV)^2} \frac{1}{(s + \lambda)} \tag{12.6}$$

In the double limit $k, s \to 0$, which is the long time, large distance limit, $P(k, s)$ takes on the familiar form,

$$P(k, s) \approx \frac{\lambda}{(s\lambda) + (kV)^2} \tag{12.7}$$

which transforms back in the Gaussian,

$$P(x, t) \approx \frac{1}{\sqrt{4\pi Dt}} \exp\left(-\frac{x^2}{4Dt}\right) \tag{12.8}$$

with $D = V^2/\lambda$.

12.2. The Telegrapher's Equation

Let us now do the exact inverse Fourier transform of $P(k, s)$ to have

$$P(x, s) = \frac{[(s + \lambda)^2 + (s^2 + \lambda s)]}{2(s + \lambda)\sqrt{s^2 + \lambda s}} \exp\left(-\left|\frac{x}{V}\right|\sqrt{s^2 + \lambda s}\right)$$

and we now take the small s limit

$$\lim_{s \to 0} P(x, s) \approx \frac{\lambda \exp\left(-\left|\frac{x}{V}\right|\sqrt{s^2 + \lambda s}\right)}{2\sqrt{s^2 + \lambda s}} \tag{12.9}$$

writing,

$$\sqrt{s^2 + \lambda s} = \sqrt{\left(s + \frac{\lambda}{2}\right)^2 - \frac{\lambda^2}{4}}$$

and using the identity

$$\mathcal{L}^{-1} \frac{\exp(-k\sqrt{s^2 - a^2})}{\sqrt{s^2 - a^2}} = I_0(a\sqrt{t^2 - k^2})\Theta(t - |k|) \tag{12.10}$$

where $\Theta(t - |k|) = 1$ if $t > |k|$ and zero, otherwise, and I_0 is a modified Bessel function of the first kind, then

$$P(x, t) = \exp(-\frac{\lambda t}{2})I_0\left(\frac{\lambda}{2}\right)\sqrt{t^2 - \frac{x^2}{V^2}}\Theta(Vt - |x|) \tag{12.11}$$

Now $I_0(z) \approx \exp(z)/\sqrt{2\pi z}$ for large z, and here

$$z = \frac{\lambda t}{2}\sqrt{1 - \frac{x^2}{V^2 t^2}} \approx \frac{\lambda t}{2}\left(1 - \frac{x^2}{2V^2 t^2}\right)$$

and asymptotically for long times,

$$P(x, t) \approx \frac{1}{\sqrt{\pi \lambda t}} \exp\left(-\frac{x^2}{4V^2 t/\lambda}\right)\Theta(Vt - |x|) \tag{12.12}$$

the probability looks like the standard diffusion result, except now in this more precise calculation, there is a cut-off at $|x| = Vt$. The diffusion equation was not physical in that the probability to be far away at early times was non-zero. The telegrapher's equation fixes that by having the probability front move out with a velocity V.

In the CTRW formalism the term $\frac{1-\psi(s)}{s}$ accounts for the walker remaining at a site x for a time $t \geq \tau$ when the walker reached site x at time $t-\tau$. But, with the walkers always moving with a velocity V, there are turning points, but no waiting time at a site. The probability to be at a site at time t can now be accounted for by reaching the site being a turning point a time t, or by passing over the site at time and reaching a turning point later on. The CTRW formula, with its waiting times between jumps, is not the correct formalism for a walker always moving with a velocity and no stopping.

12.3.   Albers–Radons Flights

Albers and Radons (*Phys. Rev. Lett.* **120**, 104501 (2018)) showed the correct analysis for a Lévy walk that is constantly in motion with the distance between turning points governed by a probability distribution also allowing jumps of different lengths having different velocities. We present their part of their analysis, and for simplicity stick to 1D. They treat the case for any dimension and arbitrary exponents governing flight times and distances. Start with $\psi(t)$ being, as before, the conditional probability of a jump of x given that it was of duration t,

$$\psi(x|t) = \frac{1}{2}\delta(|x| - Vt)\psi(t) \qquad (12.13)$$

where V will depend on the flight duration, so if $x = Vt \propto t^\beta$, then $V \propto t^{\beta-1}$.

We will look at the case $\psi(t) \propto t^{-1-\alpha}$. For the mean squared displacement, we average over jumps of duration τ governed by $\psi(\tau)$ having a velocity $V(\tau)$ and a displacement at time t of $x(t) = V(\tau)t$. For just the first jump mean squared displacement, accounting for flights that are of duration greater than time t,

$$\langle x^2(t) \rangle_{\text{single jump}} = \int_t^\infty |V(\tau)t|^2 \, \psi(\tau) d\tau$$

$$\sim t^2 \int_t^\infty \tau^{2\beta-2-1-\alpha} d\tau \qquad (12.14)$$

which diverges if $2\beta \geq \alpha + 2$. So, a very different result than the CTRW equation which has a probability of being at a site at an earlier time $t - \tau$ and not leaving by a time τ. However, we will show below that when the mean jump lengths are finite then the CTRW would asymptotically be correct. The more interesting cases are found in the Albers–Radons work.

In 1D, the probability for a jump of displacement $x + dx$ at a time t, taking into account all jumps of duration τ for all $\tau > t$ so the walker lands or passes over x at time t,

$$\Omega(x, t) dx = \frac{1}{2} \int_t^\infty \delta(|x| - \tau^{\beta-1}t)\psi(\tau) d\tau dx \qquad (12.15)$$

and the probability that the walker is somewhere at time is unity so

$$\int_{-\infty}^\infty \Omega(x, t) dx = 1$$

and this holds in all dimensions.

For the calculation for many jumps, as before we can write, in (Fourier k, Laplace s) space

$$P(k,s) = \Omega(k,s) + \Omega(k,s)\psi(k,s) + \Omega(k,s)\psi^2(k,s) + \cdots$$
$$= \frac{\Omega(k,s)}{1 - \psi(k,s)} \quad\quad (12.16)$$

First, let's look asymptotically at the case when all space and time moments are finite with constant velocity $V = 1$ with $\beta = 1$.

$$\Omega(k,t) = \frac{1}{2}\int_{-\infty}^{\infty} e^{ikx}\int_{t}^{\infty}\delta(|x| - \tau^{\beta-1}t)\psi(\tau)d\tau dx$$
$$= \cos(kt)\int_{t}^{\infty}\psi(\tau)d\tau \quad\quad (12.17)$$

For $k \to 0$, $\Omega(k,t) = \int_{t}^{\infty}\psi(\tau)d\tau$.

So in Laplace space $\Omega(k = 0, s) = \frac{1-\psi(s)}{s}$, the waiting time term familiar from the CTRW, thus we asymptotically recover the CTRW formula as k goes to zero

$$P(k,s) = \frac{1}{1 - \psi(k,s)}\frac{1 - \psi(s)}{s}$$

which gives the standard Brownian-type result. This won't be the case, in general, when infinite moments are involved.

Let us consider the case of constant velocity $V = 1$ ($\beta = 1$) and infinite mean jump times, $\psi(t) \sim t^{-1-\alpha}$ for large t with $\alpha < 1$ to find for an infinite single jump that takes an infinite amount of time, but at time t, x will only travel Vt. The mean displacement is zero when there is no bias, so let's work the second moment. The mean squared displacement in Laplace s-space is,

$$\langle x^2(s) \rangle = -\frac{\partial^2}{\partial k^2} P(k, s)|_{k=0}$$

For this case, the dominant term for computing $\langle x^2(S) \rangle$ will be

$$\langle x^2(s) \rangle = -\lim_{K=0} \frac{\partial^2}{\partial k^2} \frac{\Omega(k, s)}{1 - \psi(k, s)} \tag{12.18}$$

Next,

$$\Omega(k, t) = \frac{1}{2} \int_{-\infty}^{\infty} e^{ikx} \int_{t}^{\infty} \delta(|x| - Vt)\psi(\tau) d\tau dx$$

$$= \cos(kt) \int_{t}^{\infty} \psi(\tau) d\tau$$

yielding,

$$-\frac{\partial^2}{\partial k^2} \Omega(k, t) \Big|_{k=0} = t^2 \int_{t}^{\infty} \psi(\tau) d\tau$$

Taking the Laplace transform,

$$-\frac{\partial^2}{\partial k^2} \Omega(k, s) \Big|_{k=0} = \frac{\partial^2}{\partial s^2} \int_{0}^{\infty} e^{-st} \int_{t}^{\infty} \psi(\tau) d\tau$$

$$= \frac{\partial^2}{\partial s^2} \frac{1 - \psi(s)}{s} \sim \frac{1}{s^{3-\alpha}}$$

and since

$$1 - \psi(s) \sim s^{\alpha}$$

then

$$\langle x^2(s) \rangle \sim \frac{1}{s^3}$$

$$\langle x^2(t) \rangle \sim t^2 \tag{12.19}$$

which for $\beta = 1$ and $\alpha < 1$ follows the inequality $2\beta < \alpha + 2$ for a non-divergent second moment. This is the ballistic result and the greatest time dependence for when the CTRW model holds asymptotically.

For all the more interesting possible outcomes for ranges of α and β see the Albers–Radons, *Phys. Rev. Lett.*, Vol. 120, p. 104501 (2018).

CHAPTER 13

Random Walks With Internal States

13.1. Matrix Formalism

We can generalize the CTRW to include internal states at each site. These might represent different configurations, a memory of what path was taken to the site, accounting for if the site is occupied by non-radioactive or radioactive particle, possible ways to fill a vacancy, and any other condition or situation of which you can think. One now asks questions like what is $P_i(l, t)$ the probability of being at site l at time t in state i, or calculating $P_{i,j}(l, t)$ the same probability, but specifically if starting in state j.

The more general CTRW equation looks the same,

$$Q(l,t) - \sum_{l'=-\infty}^{\infty} \int_0^t Q(l - l', t - \tau)p(l')\psi(\tau)d\tau = \delta_{l,0}\delta(\tau)$$

(13.1)

but now the variables are matrices. In Fourier–Laplace space,

$$Q(k,s)[1 - p(k)\psi(s)] = 1,$$

but we do not simply divide by $1-p(k)\psi(s)$, but need to multiply each side of the equation by its inverse matrix. Then,

$$P(k, s) = [1 - p(k)\psi(s)]^{-1}\frac{1 - \psi(s)}{s} \qquad (13.2)$$

13.2.   Dimer Diffusion

Let's work out an example to make this more clear. The example comes from field ion microscopy where adatoms (atoms adsorbed on a surface) were imaged moving randomly on the surface of a crystal face in a channel. Two atoms in neighboring channels were observed to form a dimer and moved together as a dimer cluster alternating jumps. The atoms could be directly across from each other or in a stretched configuration where one was advanced by a lattice spacing. Moving from the stretched configuration back to the straight configuration occurred with a rate β and moving back to the stretch configuration was with rate α.

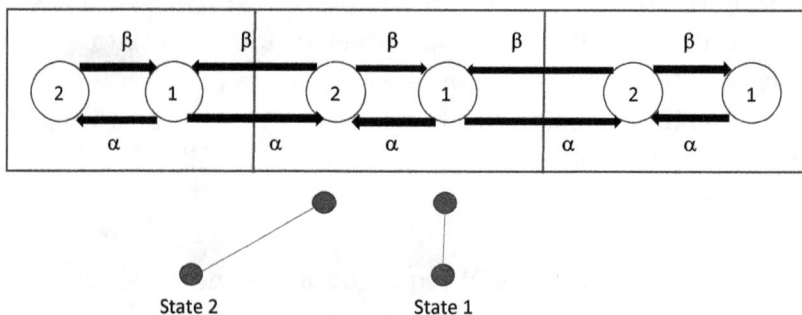

State 2   State 1

The lattice spacing is $L = 1$.

The matrix for jumps between the configurational sites is,

$$\psi(t)p(l) = \begin{pmatrix} 0 & \frac{1}{2}\beta \exp(-\beta t)(\delta_{l,0} + \delta_{l,1}) \\ \frac{1}{2}\alpha \exp(-\alpha t)(\delta_{l,0} + \delta_{l,-1}) & 0 \end{pmatrix}$$

$$(13.3)$$

with

$$\Phi(s) = \frac{1 - \psi(s)}{s} = \begin{pmatrix} \frac{1}{\alpha+s} & 0 \\ 0 & \frac{1}{\beta+s} \end{pmatrix} \tag{13.4}$$

and

$$[1 - p(k)\psi(s)]^{-1} = \frac{1}{\Delta} \begin{pmatrix} 1 & \frac{\beta}{(\beta+s)}(1 + \exp(ikL)) \\ \frac{\alpha}{(\alpha+s)}(1 + \exp(-ikL)) & 1 \end{pmatrix}$$

where we consider here a general size L and Δ is the determinant of the matrix,

$$\Delta = \left[1 - \frac{1}{2} \frac{\alpha\beta}{(\alpha+s)(\beta+s)}(1 + \cos(kL)) \right]$$

giving,

$$P(k, s) = [1 - p(k)\psi(s)]^{-1}\Phi(s)$$

$$= \frac{1}{\Delta} \begin{pmatrix} \frac{1}{\alpha+s} & \frac{\beta}{(\beta+s)}(1 + \exp(ikL)) \\ \frac{\alpha}{(\alpha+s)}(1 + \exp(-ikL)) & \frac{1}{\beta+s} \end{pmatrix}$$

We will look at the long-time limit corresponding to s going to zero for the mean squared displacement when there is no bias, and we will average over the initial internal states j whose probabilities are p_j and we sum over the final internals states i,

$$\langle l^2(t) \rangle = -\mathcal{L}^{-1} \lim_{k=0} \frac{\partial^2}{\partial k^2} \sum_i \sum_j P_{i,j}(k, s)p_j \tag{13.5}$$

For exponential kinetics, it won't make a difference in the long run if starting in state 1 or state 2, so for $t \to \infty$, $P_{1,1}(t) = P_{1,2}(t)$, and $P_{2,2}(t) = P_{2,1}(t)$, with $p_1 + p_2 = 1$, so asymptotically

$$\sum_i \sum_j P_{i,j}(l,t)p_j = P_{1,1}(l,t) + P_{2,2}(l,t)$$

which means that we only need to work with the diagonal matrix elements (the Trace). First calculate,

$$\frac{\partial}{\partial k}\frac{1}{\Delta} = \frac{-\frac{L}{2}\frac{\alpha\beta}{(\alpha+s)(\beta+s)}\sin(kL)}{\left[1 - \frac{1}{2}\frac{\alpha\beta}{(\alpha+s)(\beta+s)}(1 + \cos(kL))\right]^2}$$

and for $k = 0$ and $s \to 0$,

$$-\frac{\partial^2}{\partial k^2}\frac{1}{\Delta} = \frac{\frac{L^2}{2}\frac{\alpha\beta}{(\alpha+s)(\beta+s)}}{\left[1 - \frac{\alpha\beta}{(\alpha+s)(\beta+s)}\right]^2} \to \frac{L^2}{2}\left(\frac{\alpha\beta}{\alpha+\beta}\right)^2$$

The trace of the $P(k,s)$ matrix for $s = 0$ is $\frac{\alpha+\beta}{\alpha\beta}$ when multiplied by the above derives the asymptotic result,

$$\langle l^2(t)\rangle = \frac{L^2}{2}\frac{\alpha\beta}{\alpha+\beta} \tag{13.6}$$

If $\alpha \gg \beta$ so β is the rate limiting step, then $\langle l^2(t)\rangle = \frac{L^2}{2}\beta$. The formalism will work for any finite number of internal states.

The question in the dimer case was to determine the activation energy from images of the dimer being in the straight or stretch state to determine the occupation probabilities experimentally. Assume the steps α and β follow the Arrhenius law,

$$\alpha = \nu \exp(-\Delta_\alpha/kT)$$

and similarly, for β. From the images of the dimer one can calculate the number of pictures N seeing the dimer in state 1 and the number M in state 2. Then the ratio $\frac{\alpha}{\beta} = \frac{M}{N}$, with M/N a known experimental quantity, so $\langle l^2(t)\rangle$ can be written as $\frac{L^2}{2}\frac{\alpha}{M/N+1}$ and a semilog plot of $\langle l^2(t)\rangle$ vs. $1/kT$ would give a slope of value equal to $-\Delta_\alpha$ and similarly a semilog plot of $\frac{L^2}{2}\frac{\beta}{N/M+1}$ would a slope of value $-\Delta_\beta$.

13.3. Memory: Permanent or Fading

For random walks with internal states where transitions among states are time dependent, we need the CTRW to track time rather than just counting the number of steps.

An example is that the internal states can hold the memory of past steps of a random walker. We'll look at the case of remembering if a walker came to a site, in 1D, from the right or left and this can influence if the walker continues in the same direction or reverses direction. A walker continues jumping in the same direction with probability p and reverses direction with probability $q = 1 - p$, and all jumps occur with a rate λ.

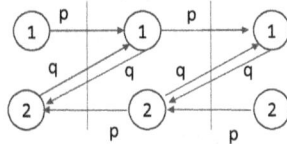

For this case,

$$\psi(t)p(l) = \lambda \exp(-\lambda t) \begin{pmatrix} p\delta_{l,1} & q\delta_{l,1} \\ q\delta_{l,-1} & p\delta_{l,-1} \end{pmatrix} \qquad (13.7)$$

$$\Phi(s) = \frac{1 - \psi(s)}{s} = \frac{1}{\lambda + s}\begin{pmatrix} 1 & 0 \\ 0 & 1 \end{pmatrix} \qquad (13.8)$$

Performing the matrix inversion

$$Q(k, s) = [1 - p(k)\psi(s)]^{-1} = \frac{M(k, s)}{\Delta}$$

$$= \frac{1}{\Delta}\begin{pmatrix} 1 - \frac{\lambda}{\lambda+s}pe^{-ik} & -\frac{\lambda}{\lambda+s}qe^{ik} \\ -\frac{\lambda}{\lambda+s}qe^{-ik} & 1 - \frac{\lambda}{\lambda+s}pe^{ik} \end{pmatrix}$$

where Δ is the determinant and

$$\Delta = 1 - \frac{\lambda}{\lambda + s} 2p \cos(kl) + \left(\frac{\lambda}{\lambda + s}\right)^2 (p - q) \qquad (13.9)$$

as $k, s \to 0$,

$$\Delta(s) \sim \frac{2(1 - p)}{\lambda} s + o(s^2) \qquad (13.10)$$

When calculating the mean squared displacement, taking the derivative of $\frac{1}{\Delta}$ we get a $\frac{1}{\Delta^2}$ term which will go as $\frac{1}{s^2}$ and when inverse Laplace transforming will give $\langle l^2(t) \rangle \sim t$, the diffusive-type result, but a complete analysis will provide an interesting coefficient. For the mean squared displacement,

$$\langle l^2(t) \rangle = - \lim_{k=0} \mathcal{L}^{-1} \frac{\partial^2 P(k, s)}{\partial k^2}$$

as the initial state being 1 or 2 won't affect the asymptotic behavior we only need calculate the second moment of $P = P_{11} + P_{22}$ where in Fourier–Laplace space this is $\frac{\lambda}{\lambda+s}$ Trace $M(k, s)$, the dominant term for the mean squared displacement will be in the $k = 0$, $s \to 0$ limit,

$$-\frac{1}{\Delta^2} \frac{\partial^2 \Delta}{\partial k^2} \frac{1}{\lambda + s} \text{Trace } M(k, s)$$

$$\frac{\partial^2 \Delta}{\partial k^2} = -p \frac{2\lambda l^2}{\lambda + s} \cos(kl)$$

and

$$\lim_{k=0, s=0} \text{Trace } M(k, s) = 2(1 - p)$$

yielding,

$$\langle l^2(t) \rangle = l^2 \frac{p}{q} \lambda t \qquad (13.11)$$

If $p = q = 1/2$ we get the regular random walk result. If $p = 1$, $q = 0$, that would be ballistic motion in one direction with a t^2

behavior and would be needing a different analysis. For $p > 1/2$ this would enhance the diffusion by making, on average, longer consecutive jumps in the same direction, and $q > 1/2$ would have the walker reverse direction often generating a reduced diffusion.

One can complicate this model by having p being time dependent, for example,

$$p(t) = 1 - q(t) = \frac{1}{2} \pm B \exp(-\beta t), \quad 0 < B < \frac{1}{2}$$

With the plus sign the walker starts with probability $1/2 + B$ to continue in the same direction, but this relaxes to probability of $1/2$ the longer the waiting time for the jump occurs. With the minus sign the walker has probability of $1/2 + B$ to change direction and this relaxes to $1/2$ the longer the waiting time before the next jump. This memory introduces a new time parameter β, but the analysis follows the same steps as before.

Equation (13.7) generalizes to,

$$\psi(t)p(l) = \lambda \exp(-\lambda t) \begin{pmatrix} p(t)\delta_{l,1} & q(t)\delta_{l,1} \\ q(t)\delta_{l,-1} & p(t)\delta_{l,-1} \end{pmatrix}$$

and the same steps can be followed to calculate random walk properties, such as the mean squared displacement.

In the theory of superionic conductors, the $+$ sign with $p(t)$ is called a caterpillar mechanism and the minus sign the bounce back mechanism. This reflects the relaxation or excitation of a lattice following the jump of an ion. Rather than calculating the mean squared displacement one calculates the frequency-dependent diffusion constant which has real and imaginary parts. There will be a rate λ for jumping and the rate β for remembering the direction of the last jump. The mean squared displacement will now have transient terms in the diffusion constant that will determine frequency terms in the frequency-dependent frequency constant. For example, the bounce back mechanism will enhance the ac conductivity. I wrote about this in more detail in *Solid State Comm.* Vol. 32 p. 1207 in 1979. My paper was submitted through the editor, Alex Maraduddin, who was a Montroll postdoc and later president of UC Irvine.

Other examples of random walks with internal states include vacancy motion in a solid where different atoms compete to fill the vacancy and the vacancy would move to a variety of locations; cyclic reversible reactions with several states represented as internal states on a lattice, where completion results in products modeled by moving a unit cell to the right or up one unit cell depending on which product is produced (or backward if the reaction is reversed). As usual with probability, the experience is unbounded, as one can imagine newer models with nothing holding you back beyond your imagination.

CHAPTER 14

Fish and Anti-Fish and Electrons and Holes

I do not remember where I saw this puzzle, but it's a good one. There is the old puzzle of three fishermen placing their catch in a single pile and as night falls they all go to sleep. The first one to wake up in the morning separates the fish into three equal piles with one fish leftover. He throws that fish away, takes one pile and pushes the remaining two piles together and leaves. The second fisherman wakes up and repeats the procedures of the first fisherman and so on for the third fisherman. The question is what is the minimum number of fish caught to fulfill this sequence of events. The obvious answer is 25. The first fisherman makes 3 piles of 8, keeps one pile and tosses away the 25th fish. The second fisherman sees 16 fish, makes 3 piles of 5, keeps one pile and tosses away the 16th fish. The last fisherman sees 10 fish, makes 3 piles of 3, keeps one pile and tosses away the 10th fish and six fish remain. Let's turn to the equations and find that -2 fish a better minimum than 25. This is an excellent example of finding a new solution to what at first appears to be a straightforward problem. Advances can be made by someone looking at a problem in an unexpected way.

Let N be the number of fish with $N = 3x + 1$. The first fisherman takes x fish and tosses one away, leaving $2x$ fish. The second fisherman finds $2x$ fish, so $2x = 3y + 1$. He takes y fish, throws one away and leaves $2y$ fish. Next, $2y = 3z + 1$ and third fisherman takes z fish. Set $N = -2$, then $x = y = z = -1$. The first fisherman sees $N = -2$ anti-fish and a fish anti-fish pair (and the pair cancels as $1 - 1 = 0$). He makes 3 piles each with one anti-fish and tosses away the fish and takes home an anti-fish, leaving behind 2 anti-fish. The second fisherman sees 2 anti-fish and a fish anti-fish pair, and so on for the third fisherman, and so on, no matter how many fishermen you want to consider.

In semiconductor physics, it is quite the norm to promote an electron from the valence band to the conduction band, leaving behind a hole, with the electron–hole pair playing the role of our fish anti-fish pair. For anomalous diffusion, we'll focus on an experiment with hole transport.

CHAPTER 15

Harvey Scher: An Appreciation

15.1. An Accidental Short Meeting with Elliott Montroll

It was a Saturday morning in the Fall of 1973. I was at the university working in my cubicle (one of seven in that room) when I heard Elliott Montroll pass by into his office further down the hall. This was unusual as he typically only came in once a week, usually on Monday. As an Einstein Professor paid by New York State, he was free from teaching and administrative duties, and either worked from home or was consulting. One of his consulting positions was with the Xerox Research Center in nearby Webster, NY close to Lake Ontario. I came out of my cubicle to his office to say hello. He was packing some material into a briefcase and said, he was leaving on sabbatical to Switzerland. That was a shock. I asked if there was anything he would like me to do while he was away. Back then there was no email or Face Time to stay in touch, mail was slow and overseas calls were expensive. He thought for a moment and said that he was working on a problem with Harvey Scher at Xerox and I should introduce myself to Harvey, who was a nice guy. Montroll then left. I knew of Scher's beautiful work on ac conductivity in amorphous semi-conductors and went ahead and called him. He was gracious and we set up a meeting. At Xerox I was impressed that on his floor, the hall along the wall, had a row of Xerox machines. At the university, we only had one in the Department Chair's side office and you had to sign for every page copied. Here at Xerox the people were free to copy as much as their hearts desired. Harvey explained to me what he was working on and what they needed to overcome to complete their work. It was an explanation of photoconductivity in amorphous films used in xerography. We will discuss that in Sec. 15.3. If I had not come in to work on that Saturday morning, I would have missed Montroll and maybe never got to work with Harvey Scher. My life would have been very different as I was still fishing around for a thesis topic and the work with Harvey produced most of my thesis content over the next weekend. The work with Harvey eventually became a joint cover article in the January 1991 issue of Physics Today, along with John Bendler from GE as we included a theory of stretched exponential relaxation which was an important part of the glass transition. More about that in Chapter 16. Now onto the work of Scher and Lax on ac conductivity.

15.2.   Scher–Lax AC Conduction in Amorphous Semiconductors

Harvey Scher was a young physicist who left Bell Labs to join the new Xerox Research Labs in Webster, New York. In the early 1970s at Xerox he was working with two giants in physics, Mel Lax from CCNY and Bell Labs, and Elliott Montroll at the nearby University of Rochester. It turned out that his 1973 paper with Lax is Lax's second most cited paper, and his 1975 paper with Montroll is Montroll's most cited paper.

So in retrospect, Scher was the giant in these collaborations. Harvey's work created the field of anomalous diffusion, which in its many guises has been the topic of numerous conferences and thousands of publications. His early work on transport in amorphous materials opened a new approach to transport of water and chemicals in water tables. He is kind, generous, and brilliant. In one way, Harvey is old fashioned in that he publishes at a slow pace with well researched important papers. He has no interest in the modern game of maximizing the number of publications and coauthors and he shies away from the PR game. His large number of citations points to his importance and international standing. More about Harvey can be found in his Festschrift for his 60th birthday in the *Journal of Physical Chemistry*, Vol. 104, number 16, April 27, 2000 where in the preface Yossi Klafter and myself wrote that Harvey only publishes when "something important and valuable has been accomplished".

In 1973 Harvey Scher at Xerox and Melvin Lax published two papers "Stochastic Transport in a Disordered Solid. I. Theory" followed by II. "Impurity Conduction" devoted to ac conductivity in amorphous semiconductors. Their model involved the Montroll–Weiss continuous time random walk, a paper previously basically ignored. This work initiated a new direction of research involving scale invariant processes for non-equilibrium systems. While equilibrium phase transitions have fixed critical exponents, derived from correlation length divergences with decreasing temperature. Scher's work on non-equilibrium transport systems had exponents that varied continuously with parameters, such as temperature, and diverging waiting times between events. The equilibrium phase transition theories were blessed with a Nobel Prize, and the non-equilibrium theories, not yet. Scher followed this work with joint work with Elliott Montroll on photocurrents in thin-film materials used as xerographic films in copying machines. In this latter work the mean waiting time distribution was considered to be infinite, or longer than the time duration of the experiment. We will discuss later the consequences of random variables with infinite mean values for calculating transport properties. For example, this will have the effect of charge mobility in amorphous solids to have a dependence on external factors and not just be an intrinsic number. Scher described the process as dispersive transport and today it and similar processes are called anomalous transport, fractal time transport, strange kinetics or other nomenclatures. The probability distributions of random variables having infinite moments are connected to Lévy's 1920s work on stable distributions. These are sometimes called heavy tail or fat tail distributions.

Today there are thousands of papers and several conferences on anomalous transport. There is an expanding list of topics on anomalous transport including, the glass transition, hopping transport in amorphous semiconductors, random walks on fractal structures, blinking quantum dots, cold atoms in optical lattices, anomalous diffusion in a plasma, tracers trapped in vortices in rotating Couette flows, solute transport in water tables, transport in crowded biological cells, animal foraging behaviors, and strange kinetics in Hamiltonian systems. Techniques can involve fractional Langevin equations, fractional Fokker–Planck equations, fractional random walk master equations with fractal behavior in space, time, space–time, momentum or other variables. We will only discuss random walk equations here. Nevertheless, the literature is vast on the other methods. Check out the 2008 book, *Anomalous Transport: Foundations and Applications*, edited by Klages, Radons, and Sokolov, for many physics examples. Some more recent books can be found on Amazon searching under anomalous diffusion.

The Montroll–Weiss 1965 paper "Random Walks on Lattices II" had a section on continuous time random walks with a waiting time distribution between jumps. While the formalism was general, they only considered processes when the mean time between jumps was finite. And why not? Would infinite mean jump times make sense in a physics problem? Things were about to change. If the mean was finite that would set a characteristic scale. An infinite first moment means that there is no characteristic size which implies scale invariance. This is the hallmark of fractals. L. F. Richardson looking at the Spanish–Portuguese border, showed that measurements of a boundary depended on the minimum measuring size and Mandelbrot explored this in his paper, "How Long is Coast of Britain". No doubt, ideas were stirring, from diverging correlation lengths in phase transitions, cluster sizes in percolation near a critical probability, to infinite self-energies in Anderson localization, and a vast set of examples captured in Mandelbrot's remarkable book, *The Fractal Geometry of Nature*. For non-equilibrium phenomena, the Scher–Lax and the latter Scher–Montroll paper originated a new field of physics of anomalous transport.

The use of amorphous films in Xerox machines spurred the interest in amorphous materials. Scher and Lax were interested in impurity conduction in n-type compensated semiconductors. These are doped with more donors than acceptors. Each acceptor ionizes one donor leaving a spatial and energetic distribution of neutral and ionized donor sites. At low temperatures, electrons do not significantly transition to the conduction band, but hop among traps with the trap times much longer than the flight times between traps. Conduction occurs with an electron hop between a neutral donor to an ionized donor. Scher and Lax employed the CTRW to describe this hopping motion. The experiments found that the ac diffusion constant $D(\omega)$ of electrons varied as ω^s with s between 0.7 and 0.9 in the frequency range of 10^2–10^5 Hz.

The start of the collaboration was an equation that Scher saw on Lax's desk when they overlapped at Bell Labs,

$$\mathrm{Re}D(\omega) = \int_0^\infty \cos(\omega t)\langle \nu(t)(0)\rangle dt \tag{15.1}$$

which can be rewritten in terms of positions as,

$$D(\omega) = -\frac{1}{6}\omega^2 \int_0^\infty e^{i\omega t}\langle [r(t) - r(0)]^2 dt \tag{15.2}$$

where r is generally a vector. One approach would be to start with a quantum Hamiltonian to calculate a propagator $|\langle r|\exp(-iHt)|0\rangle|^2$, but Scher and Lax went classical with the Montroll–Weiss hopping model between localized band tail sites. Without a bias, and in 1D with $x(0) = 0$, we have from Chapter 10 that

$$\langle x^2(t)\rangle = \bar{x}^2 \mathcal{L}^{-1}\left\{ \frac{\psi(s)}{s(1 - \psi(s))} \right\} \tag{15.3}$$

For long times $\psi(t) \sim \frac{1}{t^{1+\beta}}$, with $\beta < 1$, has its Laplace transform for small s behaving as

$$\psi(s) \sim 1 - s^\beta$$

yielding for long times,

$$\langle x^2(t)\rangle \sim \bar{x}^2 \mathcal{L}^{-1}\frac{1}{s^{1+\beta}} \tag{15.4}$$

Switching notation from s to ω, the real part of $D(\omega)$ in Eq. (15.2) for small ω is,

$$D(\omega) \sim \frac{1}{6}\omega^2 \frac{1}{\omega^{1+\beta}} \propto \omega^{1-\beta} \tag{15.5}$$

the behavior seen in the impurity conduction experiments. Scher and Lax actually did something a little bit different for $\psi(t)$.

They used for the waiting time probability,

$$\psi(t) = \lambda(\lambda t)^\nu e^{-\lambda t}/\Gamma(\nu + 1) \tag{15.6}$$

with a mean jump time of $\bar{t} = \frac{\nu+1}{\lambda}$ and Laplace transform,

$$\psi(i\omega) = \frac{1}{\left(1 + |\frac{i\omega}{\lambda}|\right)^{1+\nu}}$$

to find in the limit $|\frac{i\omega}{\lambda}| \gg 1$ that

$$D(\omega) \sim (\omega/\lambda)^{1-\bar{t}\lambda} e^{-i\pi/2(1-\bar{t}\lambda)} \tag{15.7}$$

Setting $1 - \bar{t}\lambda = s$, then

$$\mathrm{Re}\,D(\omega) \sim \omega^s \tag{15.8}$$

They find $\bar{t}\lambda$ in the range of 0.1–0.3 with $\nu = -0.9$ and $\lambda = 0.03$ to get a probability distribution with sufficient weight for times greater than $\bar{t}$, but with a finite mean time. Not a transport model, but averaging over relaxation times. When $\omega = \mathbf{0}$ there is only a dc current and a transport model is needed. A dc current depends on if there is a percolation pathway and that will depend on the doping level in the semiconductor. This was not part of the original Scher–Lax paper, but this addition was addressed later by Odagaki and Lax using the coherent medium approximation method. Odagaki was a brilliant postdoc with Mel Lax at CCNY. He has continued to create important works till today in his native Japan. The Odagaki–Lax work can be found in the *AIP Conference Proceedings*, No. 109 (1982) edited by myself and Bruce West. That volume is from the conference proceeding *Random Walks and Their Applications in the Physical and Biological Sciences* covering some talks from my 1982 NBS conference.

15.3. Fractal Time and Anomalous Diffusion

Originally, the Gaussian arose from least squares fitting for Gauss tracking Ceres the largest object in the asteroid belt between Mars and Jupiter. Originally, for the Gaussian the mean was the most probable answer and the variance a measure of error. Physics had a mindset to find an answer, not a distribution of answers. There is the famous 1926 article "On Being the Right Size" by J. Haldane and a 1975 article "Of Atoms, Mountains, and Stars" by V. Weisskopf. Both find the right size for objects, be they atoms, animals, mountains or stars. Of course, there are the Maxwell–Boltzmann, Bose–Einstein, and Fermi–Dirac distributions, but they are well behaved. On the other hand, the Cauchy distribution (in physics called the Lorentzian) has an infinite second moment, but it usually is described by the full width at half max to ignore the long tail. In terms of spectral line shapes, R. Kubo found that colored noise changes the Lorentzian lineshape at high frequencies to a Gaussian, so theoretical efforts were made to avoid unphysical infinities.

For reactions, Arrhenius in 1889 introduced the concept of getting over an activation energy barrier to augment the idea of reactions not only depending on how many times two molecules hit each other. The Arrhenius law defined a waiting time scale t to get over a barrier Δ,

$$t = t_0 \exp\left(\frac{\Delta}{kT}\right) \tag{15.9}$$

Harvey Scher was drawn to introducing a distribution of barrier heights to describe transport in amorphous materials. Let's see how the Arrhenius law can generate a distribution of times with an infinite mean.

It may seem strange at first to think that distributions with infinite moments would leave the realm of mathematics and show up in physics, so let's start with a simple example.

From the Arrhenius law,

$$\frac{dt}{d\Delta} = \frac{t}{kT}$$

with Δ as a free energy barrier height, T the temperature and k the Boltzmann constant. A distribution of barrier heights $h(\Delta)$ induces a distribution of hopping time $\psi(t)$ connected by,

$$\psi(t)dt = h(\Delta)d\Delta$$

Choosing, $h(\Delta) = \frac{1}{\Delta_0}\exp(-\Delta/\Delta_0)$, where Δ_0 is the mean barrier height, then we can write

$$h(\Delta) = \frac{1}{\Delta_0}\left(\frac{t}{t_0}\right)^{-kT/\Delta_0}$$

giving

$$\psi(t) = \beta\frac{t_0^\beta}{t^{1+\beta}} \quad for \ t > t_0 \tag{15.10}$$

with $\beta = \frac{kT}{\Delta_0}$. When $\beta < 1$, then $\int_{t_0}^{\infty} t\psi(t)dt = \infty$. Just playing with exponentials it was easy to come up with a waiting time distribution with an infinite first moment.

We can also let t_0 be a random variable, and let's for fun introduce the math to solve that problem. This is an example of a function of random variables. We have the product of two random variables $X = t_0$ and $Y = \exp\left(\frac{\Delta}{kT}\right)$. In general, for two random variables X and Y with $Z = XY$ their probability distributions $p(z), f(x), g(y)$ are connected by

$$p(z) = \int_{-\infty}^{\infty} f\left(\frac{z}{y}\right)g(y)\frac{dy}{y} \tag{15.11}$$

The $1/y$ arises from the Jacobian $J = \partial(x,y)/\partial(z,y)$ to transform from (x,y) to (z,y) These joint probabilities are related by

$$P(z,y)dzdy = M(x,y)dxdy$$

where $M(z,y) = f\left(\frac{z}{y}\right)g(y)$ and $p(z) = \int_{-\infty}^{\infty} P(z,y)dy$. We can now write, with $x = z/y$

$$P(z,y) = M(x,y)J = M(x,y)\begin{vmatrix} \partial x/\partial z & \partial x/\partial y \\ \partial y/\partial z & \partial y/\partial y \end{vmatrix}$$

$$= M(x,y)\begin{vmatrix} 1/y & -z/y^2 \\ 0 & 1 \end{vmatrix}$$

$$= M(z/y,y)/y$$

and integrating over y gives the formula (15.11) for $p(z)$.

In our example, $X = t_0$ and $Y = \exp(\Delta/kT)$ and when $\Delta = 0$ then $Y = 1$. Choosing

$$f(x) = \frac{2}{\sqrt{\pi}}\exp(-x^2)$$

for the t_0 probability, and

$$g(y) = \beta\frac{1}{y^{1+\beta}}$$

generates the waiting time probability density,

$$p(t) = \frac{2\beta}{\sqrt{\pi}}\int_1^\infty \exp(-\frac{t^2}{y^2})\frac{dy}{y^{2+\beta}}$$

Let $u = \frac{1}{y^2}$ and $du = -2\frac{1}{y^3}dy$, then our integral becomes,

$$p(t) = \frac{\beta}{\sqrt{\pi}}\int_0^1 \exp(-ut^2)\frac{du}{u^{(1-\beta)/2}}$$

Writing, $q = ut^2$ yields

$$p(t) = \frac{\beta}{\sqrt{\pi}} \frac{1}{t^{1+\beta}} \int_0^{t^2} \exp(-q) q^{(\beta-1)/2} dq$$

As $t \to \infty$, the integral goes to a constant and the long-time behavior follows that when t_0 was fixed, but with t_0 a random variable there is now a short-time behavior without a minimum t_0. I never came across in the literature the Arrhenius law with both the prefactor and the barrier height as random variable, so I worked it out for fun as an example of how to treat products of random variables.

In the next chapter, Scher and Montroll use an infinite mean waiting time to play a central role in a transport model.

15.4. Scher–Montroll Photoconductivity in Xerographic Films

The Webster, New York site for the Xerox research Center was tasked with a better understanding of the Xerox process in order to improve the copying machine. The flash of light when starting the copying process was absorbed on an amorphous film (originally As_2Se_3) where there was print on the page and reflected from the white part of the paper. This creates an electrostatic image of the dark areas that fuse with negatively charged toner particles to form a printed image. An experiment studied how light induced electron–hole pairs split apart and moved across a voltage in an amorphous semiconductor film to opposite polarity electrodes. The experiment measured the hole current and found an unusual result. For a film of width L and an electric field E across the sample, and a transit time T to cross the sample, the hole mobility μ should move with a velocity $v = L/T$,

$$v = \mu E$$

or with rewriting as,

$$\frac{1}{T} = \mu \frac{E}{L}$$

Surprisingly, the result for As_2Se_3 was

$$\frac{1}{T} = \mu \left(\frac{E}{L} \right)^{2.2} \tag{15.12}$$

In the theory a non-integer exponent for a waiting time between jumps changes the dimensionality of the velocity to give the above result. Furthermore, the measured current behaved as,

$$I(t) \sim t^{-0.45} \tag{15.13}$$

and then switched over at a later time to

$$I(t) \sim t^{-1.55} \tag{15.14}$$

with the sum of the exponents equal to -2. These currents lasted for longer than anyone cared to do the experiment. These were the experimental results that Scher showed me in his office on our first meeting, plus his CTRW calculation using a waiting time distribution with an analytic Laplace transform for an iterated error function that produced a current decaying as $t^{-1/2}$. So Scher was 99% of the way of completing his model of the photocurrent, but he graciously asked me to see how to generalize the result. After the visit, over the weekend, I worked out Scher's calculations for an asymptotic algebraic waiting time density and I had a thesis topic.

If $\psi(t) \sim t^{-1-\beta}$, with $\beta < 1$, then, $\int t\psi(t)dt = \bar{t}$ diverges, and for small s, $\psi(s) \sim 1 - s^{\beta}$.

The mean current depends on the mean position r,

$$\langle I(t) \rangle = \frac{d}{dt} \langle r(t) \rangle$$

In Chapter 10, we calculated that,

$$\langle r(t) \rangle = \sum_r r P(r,t) = \bar{r} \mathcal{L}^{-1} \frac{1}{s} \frac{1}{1 - \psi(s)} \tag{15.15}$$

where $\bar{r}$ is the mean jump distance, and for our infinite moment $\psi(s)$ we find,

$$\langle r(t) \rangle = \bar{r} \mathcal{L}^{-1} \frac{1}{s} \frac{1}{s^\beta} \sim \bar{r} t^\beta \tag{15.16}$$

Note that we never calculate or directly use $\int_0^\infty t\psi(t)dt = \infty$. Let T be the transit time when the mean distance traveled is L, and let $\bar{r} \propto E$, when the average distance equals the sample size L, then

$$\langle r(T) \rangle = L = \bar{r} T^\beta \propto E T^\beta \tag{15.17}$$

rearranging,

$$\frac{1}{T} \propto \left(\frac{E}{L} \right)^{1/\beta} \tag{15.18}$$

For As_2Se_3, the data is fit by $\beta = 0.45$, so $\frac{1}{\beta} = 2.2$, solving the mobility riddle of μ not being constant, but depending on E and L. For larger L the random walkers have more opportunity to land in deeper wells to slow the current. Experiments with TNF-PVK found a $\beta = 0.8$. The current for As_2Se_3,

$$\langle I(t) \rangle = \frac{d}{dt} \langle r(t) \rangle \propto t^{-1+\beta} = t^{0.45}$$

but why does the current decrease at the transit time to a faster rate. It is because charges are lost at the opposite polarity electrode at site L. To calculate the loss of charges over time we need to calculate the first passage time probability to reach L to subtract those walkers from contributing to the current. The result will be the current transitioning to a $t^{-1-\beta}$ behavior and the early $(-1+\beta)$ and late $(-1-\beta)$ exponents adding up to -2, what I've called the "minus 2 law".

Let's do the calculation for the probability to be at a site accounting for the possibility that it might have been absorbed at the boundary at site L:

$$P(r,t) = P_0(r,t) \left[1 - \int_0^t [F(L,\tau)] d\tau \right]$$

where the subscript 0 is for the probability with no boundary at L, and F is the first passage time probability density. The integral is the probability that the walker did not encounter the absorbing barrier at an earlier time τ. The equation for the mean position,

$$\langle r(t) \rangle = \langle r(t) \rangle_0 \left[1 - \int_0^t [F(L,\tau)] d\tau \right]$$

so, at short times the same behavior $\langle r(t) \rangle \sim t^\beta$ is found for the case with no absorbing barrier.

Consider, the known (we won't derive it here) Laplace transform of the expression in the bracket of the above equation for a biased continuous time random walk in 1D, with $p = 1 - q$ the probability of jump to the right,

$$\frac{1}{s} - \frac{1}{s} F(L, \psi(s)) = \frac{1}{s} \left(1 - \left[\frac{1 - \sqrt{1 - 4pq\psi^2(s)}}{2q\psi(s)} \right]^L \right)$$

for small q and small s with $\psi(s) \sim 1 - s^\beta$, we have

$$\frac{1}{s} - \frac{1}{s} F(L, \psi(s)) \sim \frac{1}{s^{1-2\beta}}$$

giving at long asymptotic times $\langle r(t) \rangle \sim t^\beta t^{-2\beta}$ with the current

$$I(t) = \frac{d\langle r(t) \rangle}{dt} \sim \frac{1}{t^{1+\beta}}$$

and the current changing over from $\frac{1}{t^{1-\beta}} \rightarrow \frac{1}{t^{1+\beta}}$ with the exponents adding up to -2.

I wrote all of this up and mailed it to Montroll who was still in Switzerland on sabbatical. This was Scher and Montroll's research topic. Rather than being cross at his graduate student asking to publishing (the above calculations) on his own, instead Montroll was impressed. He treated me afterwards as an equal and the start of more collaborations. He said I should sent to paper to the *Journal of Statistical Physics* directly to Dr. Katja Lindenberg as the referee. Kajta had been Montroll's postdoc for one year in 1969 and then she moved to UCSD and started her 50-year career there. Katja accepted the paper and we started a long friendship. Katja created a statistical physics center at UCSD with visitors from around the world. She was always supportive of many people, including myself. Here's a copy of the first page of a letter that I sent her (no email in 1974) discussing one of her recent works on linear response theory. By the way over the years, I did send letters to well-known physicists asking for advice on topics and always got nice replies. I never worried about someone stealing my work, and my colleague and friend Arnold Mandell said, you'll know when your work is good if it's stolen. Anyway, here's my letter below with some algebraic tricks to shorten a derivation.

The transient photocurrent topic was my thesis and it seems fairly straightforward now, but then in the early 1970s employing and calculating with probabilities with infinite moments was somewhat mysterious.

THE UNIVERSITY OF ROCHESTER
RIVER CAMPUS STATION
ROCHESTER, NEW YORK 14627

DEPARTMENT OF PHYSICS
AND ASTRONOMY

Aug 22, 1994

Dear Prof. Lindenberg,

The discussions I had with Dick Bedeaux last June have led to a simplification of the derivation of the linear response for a random walker. Let us start with your equation (3.15) and use your notation (ie no subscripts) à la Bedeaux.

$$P_i(t) = -\beta \tau_i^{-1} \left[\sum_{k=0}^{\infty} M_o^k \int_0^t \phi_i(k) \, k(t-t')(M_o-1) \right] A \, P^{eq}_i \qquad (3.15)$$

$$\Delta \overline{B}(s) = B \cdot \overline{P}_i(s) = -\beta \tau_i^{-1} \langle B \, G(s)(M_o-1) A \rangle \, \overline{k}(s)$$

$$= -\beta \tau_i^{-1} \langle B \{ s - \gamma(s)(M_o-1) \}^{-1} \gamma(s) (M_o-1) A \rangle \overline{\delta}(s) \, \overline{k}(s)$$

where $G(s) = \sum_{k=0}^{\infty} M_o^k \phi_i(k) = \{1 - \gamma(s) M_o\}^{-1} = \frac{1-\gamma(s)}{s} = \{s - \gamma(s)(M_o-1)\}^{-1}$

and $\overline{\gamma}(s) = s\overline{\gamma}(s)/[1 - \overline{\gamma}(s)]$

Using $\frac{x}{s-x} = -1 + \frac{s}{s-x}$ with $x = \gamma(s)(M_o-1)$ we have

$$\boxed{\Delta \overline{B}(s) = \beta \overline{k}(s) g(s) [\langle BA \rangle - s \langle B G(s) A \rangle]} \qquad \text{your eqn (3.44)}$$

where $g(s) = \{ s \tau_i \overline{\gamma}(s) \}^{-1}$

$$\boxed{\Delta B(t) = \int_0^\infty dt' \int_0^\infty dt'' \, k(t-t') g(t'-t'') \phi_{BA}(t'')} \qquad \text{your eqn (3.36)}$$

where $\int_0^\infty \phi_{BA}(t) e^{-st} dt = \chi_{BA}(s) = \beta[\langle A(o) B(o)\rangle - s \langle A(o) B(s)\rangle]$
where the matrix elements have been rearranged.

So $$\boxed{\phi_{BA}(t) = -\beta \frac{d}{dt} \langle A(o) B(t) \rangle} \qquad \text{your eqn (3.37)}$$

by the properties of the Laplace transform of a derivative.
This covers everything, except, the discrete case which must be handled separately, $\psi(t) = \delta(t-\tau_i)$.
It is easy to see that the discrete case must be handled separately. Consider $\psi(t) = \delta(t-\tau_i)$, $\psi(s) = e^{-s\tau_i}$, and

By the way for our $\psi(t) \sim t^{-1-\beta}$, if we calculate mean fluctuation $\sqrt{\sigma^2(t)}$ per mean distance traveled $\langle r(t) \rangle$ it turns out they both have the same time dependence

$$\frac{\sigma(t)}{\langle r(t) \rangle} \propto \sqrt{\frac{2\Gamma^2(1+\beta)}{\Gamma(1+2\beta)} - 1}$$

which proves the following Gamma function inequality,

$$2\Gamma^2(1+\beta) > \Gamma(1+2\beta)$$

When I showed this Gamma function trick to Mark Kac, in the summer of 1975, during a visit in Lausanne, the French part of Switzerland, he said "pas mal" (not bad in French).

15.5. What Next and Where Next?

Elliott Montroll was away for the summer of 1975 when I finished writing up my thesis. I wrote to Elliott asking on what dates would he back in Rochester, so I could set up my PhD thesis defense hoping to match his dates with perspective committee members. This would be easiest to wait until the start of Fall classes when everyone was in residence. But I was greedy. I wanted to switch from a graduate student salary to a three times larger postdoc salary. Elliott was back for a few days in June, the rest of the committee was also in place and all went smoothly and we had a celebration back at Elliott's house afterwards. Life was grand. Then Elliott left for the rest of the summer and my graduate salary terminated, but the postdoc salary was forgotten. Not to worry, I was set to become Harvey's Scher's postdoc at Xerox in September. After the PhD defense, I left for Europe on a 45-day student Eurail Pass, all on my own money. I went to a meeting in Antwerpen where Nevill Mott asked me to explain the CTRW Scher–Montroll work. Mott's variable range hopping had a hopping rate with a novel temperature dependence. The newer CTRW work had a distribution of waiting times between jumps (where the exponent was temperature dependent) with an infinite mean so the average hopping rate was zero (to fit data the mean just needed to be long than the time of the experiment). Mott mentioned the Xerox work in his 1977 Nobel address.

By the way, Mott told this amusing story of how he decided to work in the field of solid-state physics. He said initially he worked on topics in quantum mechanics, but he was always scooped by Heisenberg, so he decided to find a different field. When he was visiting Niels Bohr and staying in a boarding house one morning the landlady accosted him and said, why can't you keep your room as neat as Heisenberg does. So even here he was lagging behind Heisenberg.

Also at that 1975 Antwerpen meeting I met Yossi Klafter and we became friends and later co-authors when Yossi finished a postdoc at MIT, under Bob Silbey, and moved to the Exxon Research Labs (then in Linden and latter in Annandale, NJ) where Yossi hired me as a consultant. (He later moved back to Tel-Aviv University eventually becoming its President.) On my first trip to Europe in 1970 I saw that city names in English did not match the local name, e.g. Rome was Roma. It is not clear to me why in the US we don't use Roma. After Antwerpen I visited a Rochester graduate, Vijay Khare, who was now a postdoc in Gottingen, the famous university town with world renown in math and physics from greats like Hilbert, Klein, Born, and Heisenberg. We took a trip to West Berlin and I was surprised by signs for two Frankfurts (au Main and Oder). We picked the right one on leaving Berlin in those days of a divided Germany. I also took a side trip to Leiden visiting Dick Bedeaux whom I met at a meeting in Sitges, Spain the year before. In Leiden, I saw the Kamerlingh Onnes lab where superconductivity was discovered. I asked Dick why so many French flags, and he politely told me they were Dutch flags, just the same color scheme, one with horizontal and one with vertical strips. Life was interesting and good.

Well, all was great, until I came back to Rochester and Harvey Scher said he could not offer the postdoc position that I had banked on. Apparently, there was internal politics at Xerox and someone blocked his postdoc plan. Now I was heading into September and three months without a salary and a lost job prospect. I went to CCNY to meet Melvin Lax to ask about a postdoc position, but Mel said Montroll was going to support me. (Years later, Oxford Press asked me to proofread Mel's book "Random Processes in Physics and Finance" as he had unexpectedly passed away). Montroll came to my rescue in 1975 (and again in 1980 when arranging a research scientist position at the University of Maryland, but that's another story). Montroll told me that he got a call from someone at Xerox offering me a postdoc position. It was the person who blocked Harvey Scher's plans to recruit me. Montroll turned down that offer for me and arranged my postdoc at the University of Rochester.

In 1977, I moved to Georgia Tech as a research scientist. I have fond memories of Georgia Tech. It was a productive time and I met many good people, but my future would not be there. At Rochester, I sat on faculty committees and there was great respect among the faculty. Faculty meetings were productive and cordial. Georgia Tech was a divisive department with some faculty at serious odds with others. In 1980, I left my research scientist position at the Georgia Tech Physics Department and spent three months at the Tel-Aviv University Chemistry Department as a guest of Joshua Jortner. My daughter was born that May in Israel and we returned to the US in August joining the University of Maryland at Montroll's invitation. It was my first salary since leaving Georgia Tech. I joined Maryland's Institute for Physical Science and Technology as a research scientist. Montroll at 65 had retired from Rochester and moved to the University of Maryland.

One of the first things Montroll told me was to introduce myself to Bob Zwanzig who was the leading theorist in the

Institute. Zwanzig gave me two pieces of advice. One, don't work with Montroll as he will get all the credit, and Two, don't work on random walks, there has got to be a better topic. The first point turned out to be correct. There is a joint paper with Montroll and myself on $1/f$ noise that Charles Townes liked. At a conference at the Naval Research Lab in Washington DC he mentioned the work as Montroll and someone. A paper by Howard Reiss from UCLA, referred to my work with Montroll on maximum entropy as Montroll and collaborators and I was the only coauthor. However, later at a JASON meeting, I was sitting at lunch table with a Navy group and a Navy Captain pointed out, see the guy at another table, that's Townes, the inventor of the laser. Then he said, he's the guy standing up, then he said he's the guy walking this way. When he reached our table, Townes said, Hi Mike, can you send me that $1/f$ paper, I can't find my copy, and my prestige in the Navy went up. On Zwanzig's second recommendation it turned out that in 1982, at NBS, I organized and ran a conference on random walks. As a young researcher I attended meetings, but never as an invited speaker. So, I came up the idea of creating my own meeting, and I choose the topic of random walks. George Weiss from NIH joined me and he arranged for NIH funding, while I through Harvey Scher arranged funds from Xerox, and Yossi Klafter arranged for funds from Exxon. The meeting was a great success with 45 invited talks and a large audience. Zwanzig gave a talk on the diffusion of a deterministic random walker bouncing through a linear chain of rooms connected by open windows. So, despite his advice he ended up working on a random walk-type problem. Zwanzig later sponsored publication of a PNAS paper, with Yossi Klafter and myself, on three approaches to the stretched exponential which was fashioned after Zwanzig's famous 1964 work on the equivalence of three approaches to generalized master equations. Yossi was at the Exxon Research

Lab in Annadale, NJ and Exxon was supposed to pay the page charges. Exxon did not pay and Zwanzig showed me the letter from the National Academy of Sciences revoking his publication privileges until the charges were covered. The joke was that his wife Francis was the editor of PNAS and signed the letter. We quickly got Exxon to rectify the situation, as he was not joking.

In 1983, I left the University of Maryland for the Office of Naval Research. I now had three children and was wary of living on soft money. I decided to learn the grant game from the inside. In 1986, Maryland offered me a position to return, but at ONR I was Head of the Physics Division and soon to be elevated to the Senior Executive Service, so I stayed with the government, and now in 2020 I'm still here.

Maryland did give me their Distinguished Postdoc Alum Award.

CHAPTER 16

The Glass Transition: The Fingerprints of Anomalous Diffusion

16.1. The Stretched Exponential Problem

Elliott Montroll, together with Robert Herman and others, won the 1959 Lanchester Prize for initiating the field of traffic science. Herman is perhaps even better known for his work with Ralph Alpher on the microwave background from the Big Bang. Alpher was eventually working at General Electric and that perhaps was how Montroll became a consultant at GE. One young researcher who took advantage of Montroll's visits was John Bendler, a polymer physicist. Bendler was brilliant, had expert knowledge in polymer physics and was naturally friendly. On my visits to GE, almost everyone we passed walking in the halls would say hello to him. His specialty was polycarbonate, a GE invention. Bendler had written computer programs to produce the molecular structure of polycarbonate with a variety of end groups. When the patent on polycarbonate expired, GE wanted to be ready with new patents on variants that perhaps had better properties for applications, such as, bulletproof glass, eyeglass lenses, car parts, bottled beverages. How these materials aged, especially in sunlight was of interest. But, the problem that Montroll brought back to me from Bendler was about dielectric relaxation. Typically, a Cole–Cole plot of the real vs. the imaginary part of the dielectric function produces a semicircle if dipole relaxation was purely exponential. Most data did not fit that representation and various expression in frequency were proposed. But then in 1970, Graham Williams and David Watts, working at the US National Bureau of Standards converted the frequency data into relaxation time data and found what became known as stretched exponential relaxation. That was what Montroll related to me, in 1983, as an outstanding problem.

The French engineer, Louis Vicat, was concerned with the sagging (mechanical relaxation) of wires holding up suspension bridges in the 1800s. Scientific studies of non-exponential relaxation, goes back at least to Rudolph Kohlrausch, who found an algebraic decay of charge in Leyden jars, a type of capacitor. Rudolph's son, Frederick Kohlrausch did experiments with Wilhelm Weber on magnetic forces experimenting with magnets that were hung on silk threads. He studied the stretching, and when the magnet was removed, the slow return of the thread to a contracted length. He fit this relaxation to the stretched exponential way back in 1863. What physicists called Williams–Watts relaxation was then called, by Bendler and myself, Kohlrausch–Williams–Watts relaxation, and finally Ray Orbach and Ralph Chamberlin at UCLA, who studied remnant magnetization, called it stretched exponential relaxation.

We start by considering a dipole relaxation function,

$$\phi(t) = \frac{\langle \mu(t)\mu(0) \rangle}{\langle \mu^2(0) \rangle}$$

where $\mu(t)$ is a dipole moment. In an experiment an electric field can line up dipoles. When the field is switched off the dipoles relax into random directions from thermal fluctuations generating a dipole–dipole relaxation that is connected to the dielectric constant by,

$$\frac{\varepsilon(\omega) - \varepsilon_\infty}{\varepsilon(0) - \varepsilon_\infty} = -\int_0^\infty \exp(i\omega t)\frac{d\phi(t)}{dt}dt = \varepsilon'(\omega) + i\varepsilon''(\omega)$$

which defines the real and imaginary parts of the dielectric constant, and ε_∞ and $\varepsilon(0)$ denote the high frequency and dc limits. Our pure exponential relaxation,

$$\phi(t) = \exp(-t/\tau)$$

yields

$$\varepsilon'(\omega) = \frac{1}{1 + \omega^2\tau^2}$$

and

$$\varepsilon''(\omega) = \frac{\omega\tau}{1 + \omega^2\tau^2}$$

Note that $\varepsilon''(\omega)$ has a loss peak at $\omega = 1/\tau$. For an exponential decay of spherical dipolar particles of radius R in a fluid of viscosity η, Peter Debye calculated, in 1913, that $\tau = 4\pi R^3\eta/kT$. For the exponential decay law, the Cole–Cole plot, $\varepsilon'(\omega)$ versus $\varepsilon''(\omega)$, produces a semicircle as,

$$(\varepsilon' - 1/2)^2 + \varepsilon''^2 = 1/4$$

The interesting phenomena was that many materials do not follow this semicircle behavior. By the way, Robert Cole was best known for this and other works in chemical physics. A paper of one of his students former Sivert Glarum, working at Bell Labs, started Bendler, Montroll, and myself on the path to a model for the stretched exponential. The other Cole brother, Kenneth Cole invented the voltage clamp which allowed Hodgkin and Huxley to understand and model sodium and potassium voltage controlled currents in neural membranes. Off topic, but I have a number of papers led by John Clay on noisy neural membrane currents studied using the Cole voltage clamp method.

Williams and Watts were able to invert a large number of frequency data to determine $\phi(t)$ and their results was,

$$\phi(t) = \exp\left[-\left(\frac{t}{\tau}\right)^{\beta}\right] \text{ with } \beta < 1 \qquad (16.1)$$

We will have a lot to say about τ later as we will tie it into the timescale of the glass transition.

John Bendler introduced us to Glarum's work on the diffusion of a defect reaching a frozen-in dipole and causing its immediate relaxation. It was a 1D model, that was later generalized to 3D. It was from these earlier works that we found the way to the stretched exponential. At this time, 1983, I was with Montroll at the University of Maryland/College Park in the Institute for Physical Science and Technology. I was the statistical physics seminar chairman. There was a seminar on the stretched exponential that was not satisfying and Zwanzig said, one more like that and you are out as chairman. But, Jan Sengers sent me one of his grad students to me to explain the seminar. I started off the discussion by saying that I wouldn't recommend the approach from the talk and I would start differently. By the end of our discussion I had worked out the theory on the blackboard. Montroll wrote up the manuscript and we published in the *Proceedings of the National Academy of Sciences*, Vol. 81, pp. 1280–1283 (1984). Sadly, Elliott Montroll had passed away at home in December 1983 with air tickets in his pocket for a flight later that day to meet with John Bendler at GE. His NIH experimental chemotherapy shrunk his tumor, but weakened the aorta allowing the seepage of blood causing congestive heart failure. A great loss of a beloved man with much left to contribute.

16.2. Stretched Times

Our theory starts with a jammed supercooled liquid where the only motion is initiated by mobile defects that encapsulate free volume. When a defect reaches a frozen dipole site it provides sufficient free volume to allow for dielectric relaxation through a dipole reorientation. For conductivity modeling, if a supercooled liquid has doped ions, then if a site has a trapped ion, the defect's free volume can allow the ion to jump to a new site. We posit two types of defects: single mobile defects and immobile defect pairs that collapse leaving little free volume. The paired collapse is reversible at higher temperatures or decreased pressure. The immobile defect pairs arise, through an attractive defect–defect pairing interaction increasing when the temperature is lowered or the pressure is increased. The theory is presented in the form of "defect" anomalous diffusion and the results are compared with experimental data as a function of temperature and with a special emphasis on pressure. In summary, the defect anomalous diffusion (DAD) model has the following assumptions based on the integrity of defects, their motion (kinetics), and their aggregation (thermodynamics) under temperature and pressure changes.

(1) All relaxation and transport occurs through mobile defects.
(2) The mobile defects are governed by slow anomalous diffusion.

(3) As temperature is lowered or pressure increases, the mobile defects progressively coalesce into pairs of immobile defects.

We begin by calculating the lifetime probability distribution for a frozen-in dipole. At time $t = 0$, we start with a concentration c of randomly placed mobile defects on a lattice and a frozen-in dipole whose location we label as the origin. The probability that the dipole has not been visited by a defect by time t, is denoted by $\Phi(t)$ and is given by,

$$\Phi(t) = \left[1 - \frac{1}{V} \sum_{\ell \neq 0} \int_0^t F(\ell, t) dt \right]^N$$

$$\Phi(t) \rightarrow \exp\left(-c \int_0^t \sum_{\ell \neq 0} F(\ell, t)\, dt \right)$$

as $N, V \rightarrow \infty$

with $N/V \rightarrow c$ (16.2)

where N is the number of defects and V is the number of lattice sites. For a defect starting at site l, $F(l, t)$ is the probability per unit time that it reaches the origin for the first time at time t and produces the relaxation. The term in the bracket is the probability that a particular defect did not reach the origin for the first time by time t, and that term is raised to the Nth power to account for none of the N defects reaching the origin for the first time by time t. In the limit of large N and V, with $N/V = c$, we arrive at a term that physically means $\exp(-c\,\text{flux}(t))$, where $\text{flux}(t)$ is the flow of defects into the origin at time t. If a defect starting at a point l at time $t = 0$ can get to the origin at time t, then a defect starting at the origin can reach site l at time t. This allows us to rewrite,

$$\Phi(t) = \exp(-cS(t)) \tag{16.3}$$

where $S(t)$ is the number of new sites a random walking defect visits in a time t. For anomalous diffusion, when the probability density for a jump to occur at time t decreases asymptotically at long times as

$$\psi(t) \approx \tau_0^\beta / t^{1+\beta} \tag{16.4}$$

(the algebraic waiting time probability density reflects the randomness of the environment that a defect encounters) this leads asymptotically to

$$S(t) \approx (t/\tau_0)^\beta,$$

where τ_0 is a constant.

And when substituted into Eq. (16.3) the following relaxation law follows:

$$\Phi(t) \approx \exp(-c \, (t/\tau_0)^\beta) \tag{16.5}$$

In this theory, the stretched exponential law arises as a probability limit distribution.

Note that in 1D, a Brownian particle has $S(t) \sim t^{1/2}$ which was Glarum's result and when treated in 3D, $S(t) \sim t$, the exponential relaxation results. So, the Brownian result in 1D looked promising, but it was just an artifact of being in 1D. Other 1D studies, such as an Ising magnet where changes only occur at boundaries of up and down spin domains, have the boundary move as a random walk and also find the square root exponential relaxation. The same behavior was found for coupled oscillators in 1D with a fluctuation moving as a 1D random walk.

The author lecturing about stretched exponential relaxation in Cargese, Corsica.

From the set of defects, we select the first one to reach the frozen dipole. This is an example of extreme statistics and the Weibel distribution. Extreme statistics finds the distribution of the maximum value or the minimum value from a set of identical random variables. The variables may have finite or infinite moments. In our case, we were looking for the minimum time from the collection of defects for one to reach the origin. For the interested reader, the Gumbel and the Frechet are other types of extreme value distributions.

16.3. Divergent Timescales

Once again, perhaps the best-known timescale in chemical physics is due to Svante Arrhenius in 1889 who introduced the notation of an activation energy, Δ. So, two molecules had a reaction rate not solely due to how often they collided, but the molecules needed to overcome an energy barrier. His equation is

$$\tau = \tau_0 \exp\left(\frac{\Delta}{kT}\right)$$

and in a high friction limit this equation was derived by Kramers in 1940. An altogether different timescale was discovered by Vogel, in 1921, on the study of the viscosity of oils at low temperatures. His equation for viscosity is,

$$\eta(T) = \eta_\infty^{(T-T_1)/(T-T_\infty)}$$

which we can rewrite as

$$\eta(T) = \eta_\infty \exp\left(\frac{T_\infty - T_1}{T - T_\infty} \ln \eta_\infty\right)$$

or

$$\eta(T) = \eta_\infty \exp\left(\frac{B}{T - T_c}\right) \tag{16.6}$$

where T_c is a critical temperature. This is known as the Vogel law and it was rediscovered by Fulcher in 1925, and Tammann and Hesse in 1926, and is variously referred to by any combination of those names. We will now derive an expression for the timescale for the glass transition that is similar to the Vogel law and also includes pressure as a variable.

As a representative sample, consider polypropylene glycol doped with Li ions which exhibits a conductivity that increases with temperature and decreases with pressure. While anomalous diffusion would not seem to play a role in a steady-state property, we will show its fingerprints on physical properties in the context of our defect anomalous diffusion (DAD) model of John Bendler, John Fontanella and myself. The fingerprint is the exponent β governing the trapping time between movements of defects that encapsulate free volume. The model incorporates both mobile free volume and the thermodynamics of free volume temperature and pressure changes. The model derives expressions for a stretched exponential relaxation and a Vogel–Fulcher-like law for a timescale. Both of these quantities have a dependence on the fingerprint exponent. The timescale connects, through the Einstein relations, the quantities of diffusion, conductivity, and viscosity. The exponent will show up in these quantities, as well as, the glass fragility index, and the positron annihilation lifetime formula.

We remark here that there are many types of glassy materials studied for their structural, optical, magnetic, electrical, and thermal properties. There is no consensus that any theoretical approach is accepted as "the" theory of the glass transition. Our approach is one of many, but it has given a successful approach to predict behaviors, such as, conductivity for ion doped polymer as a function of temperature and pressure. Our approach starts with the stretched exponential relaxation.

We derive an expression for the time scale in Eq. (16.5) and start by rewriting it as

$$\Phi(t) \approx \exp(-(t/\tau)^{\beta})$$

with

$$\tau = \tau_0 c^{-1/\beta}$$

So far, we have only considered c as the concentration of all defects, but we now separate defects into two classes, mobile and immobile. In our model as we lower the temperature, the entropy is lowered by combining defects into collapsing pairs with total lower encapsulated free volume. This is appropriate if there is an attractive force between defects. We will focus on polymeric materials that are modeled with the defect attractive force and are called fragile glass. For materials without defect coalescence are called strong glasses. S_iO_2 is a strong glass and not covered by our model for fragile glasses. We account for the disappearance of mobile defects through the coalescence of two mobile defects that upon combining their free volume is large enough to get filled in while producing a lower energy state to freeze the surrounding molecules. As the temperature is raised this is reversible and two mobile defects are released. There is a thermal equilibrium population of the mobile and immobile defect populations. The polymer molecules now are categorized into three species, mobile (when next to a defect), jammed (immobile, away from defects) and frozen (next to a collapsed defect pair). The concentration c of defects is now divided into two species, c_1 (single mobile defects) and $c - c_1$ (collapsed immobile low volume defect pairs). It is only the mobile defects that contribute to relaxation so in Eq. (16.5) c is replaced by c_1 and our task is to now develop a theory for c_1.

A site has a single defect if it is occupied by a defect with probability c and none of its z neighbors are defects which has probability $(1 - c)^z$. If defects have an average separation of d and experience an attractive interaction with other defects over a correlation length ξ then in a mean field approximation,

$$c_1 = c(1 - c)^z \text{ with } z = \left(\frac{\xi}{d}\right)^3 \tag{16.7}$$

$$\xi \propto L\sqrt{\frac{T_c}{T - T_c}} \tag{16.8}$$

At this point John Bendler and myself, we had the good fortune to join forces with John Fontanella and Mary Wintersgill, physics professors at the US Naval Academy in Annapolis, Maryland. They had been following our work and had high pressure data from their lab. They were studying the effects of pressure on the conductivity of ion-doped polymers in application to solid state batteries. A collaboration led us to many joint works, some of which are described below. Jumping ahead in the story, John Bendler eventually spent some years teaching physics at the Naval Academy, and I spent two years as the Kinnear Professor of Physics at the Naval Academy 2008–2010, while still holding down my Office of Naval Research responsibilities.

Here is how we added pressure P to our derivation of a timescale. Allowing for compressibility χ we use for the average defect separation,

$$d = d_0(1 - \chi P) \qquad (16.9)$$

(or we can use higher order polynomial expressions for average defect–defect separation). We have relied on a lattice gas mean field type phase transition for the defects at a critical temperature T_c, and L in Eq. (16.8) is a correlation length. Together Eqs. (16.7)–(16.9) provide a formula for the timescale,

$$\tau(T) = \tau_0 c^{-1/\beta} \exp\left(\frac{L^3 \ln(1-c)T_c^{3/2}}{\beta(T-T_c)^{3/2}d_0^3(1-\chi P)^3}\right), T > T_c \qquad (16.10)$$

This should be contrasted with the Vogel–Fulcher law,

$$\tau = \tau_0 \exp\left(\frac{B}{T - Tc}\right)$$

Equation (16.10) is derived from the DAD model. Note that c the anomalous diffusion exponent, appears in both the prefactor and in the exponential argument of the timescale law. Note also, that unlike the Arrhenius law, the prefactor is not connected to an attempt frequency to go over a barrier, but it is connected to the mobile defect concentration.

Also note in Eq. (16.10) $T - T_c$ is raised to the 3/2 power, the 1/2 from mean field theory and the 3 from the material being three dimensional. In Eq. (16.8) T_c is the temperature at which all of the mobile defects disappear. This actually does not happen because the percolation of rigidity occurs at a higher than glass transition temperature T_g invalidating our model of freely diffusing defects. Below T_g there is continued relaxation from single mobile defects blocked by rigid regions from coalescing into immobile pairs. The DAD model is for T above and close to T_g. In fact, there is no T_g in the model, but a $T_c < T_g$.

16.4. Fragility Index

A semilog plot of the derivative of a dynamical quantity x such as conductivity, viscosity, or relaxation time versus T/Tg is called a fragility plot and fragility which is denoted by m is defined by Austin Angell as

$$m = \left(\frac{\partial \log x}{\partial (T_g/T)} \right)_{T_g} \tag{16.11}$$

If $\log x$ vs. T_g/T produces a straight line, this is an Arrhenius behavior and the fluid is called a strong liquid. Our attractive potential between defects produces Eq. (16.10) which generates curved fragility plots. These materials are called fragile liquids as a small change in temperature around T_g produces a larger than Arrhenius change in material response, i.e., the material is fragile to temperature change. The defect diffusion model prediction for the fragility m is

$$m \propto \frac{T_c^{3/2} L^3 \ln(1 - c)}{\beta d_0^3 T_g^3 \left(1 - \frac{T_c}{T_g} \right)^{5/2}} \tag{16.12}$$

Note this predicts a $1/\beta$ dependence form, where β is the exponent in the stretched exponential relaxation that arose from defect anomalous diffusion. For the interested reader more information and a $1/\beta$ dependence can be found in semilog plot data in J. T. Bendler *et al.*, *J. Chem. Phys.*, **118**, 6713 (2003).

16.5.   Fingerprints in Conductivity, Dielectric Relaxation, and Viscosity

While we began our discussion in the context of dielectric relation. The Einstein relations let us, equally as well, study diffusion D, conductivity σ and viscosity η near the glass transition, as all depend on $\tau(T, P)$ through the equations

$$D = \frac{\ell^2}{6\tau}$$

$$\sigma = \frac{q^2 n D}{kT}$$

$$\frac{1}{\eta} = \frac{6\pi D r_0}{kT}$$

where for conductivity, q is charge and n is charge concentration.

As τ depends on β, the fingerprints of anomalous diffusion appear in all the above quantities, just as it appeared in the fragility index. At a fixed temperature, with increasing pressure defects are pushed closer together and more will coalesce so c_1, the concentration of single mobile defects decreases, making σ and D decrease and viscosity η increase. For the conductivity case of ion-doped polymeric liquids the defects facilitate the ion motion as well as dielectric relaxation. One might suppose that increased pressure would help ions jump to now closer nearby sites, but the defect mediated jump model captures the experimental behavior of a lower conductivity as fewer mobile defects (and more immobile ones) are now available to enhance ion movement. The conductivity does increase with temperature.

A semilog plot of the conductivity of Li ion-doped polypropylene glycol vs. pressure showing a decrease in conductivity as pressure is increased can be found in J. T. Bendler *et al. Phys. Rev. Lett.*, **87**, 195503-1 (2001).

16.6.   Free Volume from Positronium Lifetimes

As we have invoked the concept of free volume we now derive, within our model, the average size of a defect's free volume as a function of temperature. The experimental measurement technique is positron annihilation lifetime spectroscopy which is referred to as Positronium Lifetimes (PALS). It relies on an embedded Na^{22} nucleus undergoing beta decay which releases a positron and a gamma ray. The detection of the gamma ray starts a clock. The positron encounters an electron and forms positronium, an electron–positron "atom" in the material, either ortho (spins aligned and the case we study) or para (opposite spins). The wavefunctions of the electron and positron overlap, and in a vacuum the positronium lifetime is on the average 142 nanoseconds, i.e., the time for the matter–antimatter to annihilate and produce two gamma rays. The gamma ray pair has a lower energy detection than the gamma ray from the beta decay, and its detection stops the clock. In this way, the lifetime of the positronium is measured. In a confined space, the lifetime is shorter than in a vacuum and a calculation deduces the volume of the confining space. This calculation is used to produce the average free volume size as a function of temperature. The result is an S-shaped curve, at the top reflecting the average size of a single mobile defect and at the bottom left the smaller average size of an immobile dimer cluster. In our interpretation and theory, the curve represents an exchange of populations of single defects with immobile dimer defects, and not a continuous change in size of a defect. The data can be found in Bendler *et al.* in *Phys. Rev. E* **71**, 031508 (2005).

Let's now derive the PALS formula. At a high enough temperature, let there be N mobile single defects each with volume V_{single}. The concentration of defects is c and c_1 is the concentration of single defects, and $c_1 = c$ at high temperatures. As the temperature is lowered, the total volume of single defects is $N\frac{c_1}{c}V_{\text{single}}$ and now at lower temperatures dimers have formed with total volume

$$\left(1 - \frac{c_1}{c}\right)\frac{N}{2}V_{\text{dimer}}$$

and the total number of defects is $Nc_1/c + (1/2)N(1 - c_1/c)$, so the average volume in a PALS experiment is $V_{\text{PALS}} = $ (total defect volume)/(total number of defects)

$$V_{\text{PALS}} = \frac{N\frac{c_1}{c}V_{\text{single}} + \frac{N}{2}\left(1 - \frac{c_1}{c}\right)V_{\text{dimer}}}{N\frac{c_1}{c} + \frac{N}{2}\left(1 - \frac{c_1}{c}\right)}$$

$$= \frac{c_1 V_{\text{single}} + \frac{(c-c_1)}{2}V_{\text{dimer}}}{c_1 + \frac{(c-c_1)}{2}} \qquad (16.13)$$

Note when $c_1 = c$ that $V_{\text{PALS}} = V_{\text{single}}$ and when $c_1 = 0$ then $V_{\text{PALS}} = V_{\text{dimer}}$.

If there are N single defects at a high temperature, then at lower temperatures this converts to $N/2$ dimers. The total defect volume is $c_1 N V_{\text{single}}$ plus $(c - c_1)/2\, N V_{\text{dimer}}$, where for dimers we divided by 2 as two singles comprise one dimer. Substituting for c_1 yields the formula used to fit the PALS data,

$$V_{\text{PALS}} = \frac{V_{\text{single}}\exp\left[-\left(\frac{\beta BT_c^{3/2}}{(T-Tc)^{3/2}}\right)\right] + V_{\text{dimer}}\left\{1 - \exp\left(\frac{\beta BT_c^{3/2}}{(T-Tc)^{3/2}}\right)\right\}}{\exp\left[-\left(\frac{\beta BT_c^{3/2}}{(T-Tc)^{3/2}}\right)\right] + (1/2)\left\{1 - \exp\left(\frac{\beta BT_c^{3/2}}{(T-Tc)^{3/2}}\right)\right\}},$$
$$T > T_c \qquad (16.14)$$

and finally, here again, the anomalous diffusion exponent β appears.

16.7. Disappearance of the Fingerprints: Entropy/Enthalpy Ratio

In the last chapter, we saw the fingerprint β of anomalous diffusion on a variety of quantities connected to the glass transition. Here we show a calculation where those fingerprints cancel out. The relative effects of volume and temperature on the properties of glass-forming liquids is a topic of current interest. In order to separate these effects, the ratio of the apparent isochoric activation energy, E_V, to the isobaric activation enthalpy, E_P can be calculated within the DAD model. By definition

$$\frac{E_V}{E_p} = \left(\frac{\partial \ln \tau}{\partial T}\right)_V \bigg/ \left(\frac{\partial \ln \tau}{\partial T}\right)_P \tag{16.15}$$

From Eq. (16.10) and since we are calculating a ratio it is convenient to use

$$\ln \tau \propto \frac{T_c^{3/2}(P)}{|T - T_c(P)|^{3/2}V}$$

where the $d^3(1 - xP)^3$ is represented by a pressure-dependent volume $V(P)$. First, we calculate

$$\left(\frac{\partial \ln \tau}{\partial T}\right)_P = (-3/2)\frac{T_c^{3/2}}{|T - T_c|^{5/2}V} - \frac{T_c^{3/2}}{|T - T_c|^{3/2}V^2}\frac{\partial V}{\partial T} \tag{16.16}$$

and use the isobaric thermal expansion $\alpha_p = \frac{1}{V}\frac{\partial V}{\partial T}$ to express Eq. (16.16) as

$$\left(\frac{\partial \ln \tau}{\partial T}\right)_P = -\frac{T_c^{3/2}\left[\alpha_P + \frac{3}{2}\frac{1}{T-T_c}\right]}{|T - T_c|^{3/2} V}$$

$$= \frac{1}{V}\frac{3}{2}\frac{T_c^{3/2}}{|T - T_c|^{5/2}}\left[1 + \frac{2}{3}\alpha_P|T - T_c|\right] \tag{16.17}$$

For the constant V derivative, the term $\left(\frac{\partial T_c}{\partial T}\right)_V$ at first glance might seem to be zero or undefined, but the answer is somewhat subtle as T_c is not a constant, but is a pressure-dependent variable. We can write the following partial derivative as,

$$\left(\frac{\partial T_c}{\partial T}\right)_V = \left(\frac{\partial T_c}{\partial P}\right)_T \left(\frac{\partial P}{\partial V}\right)_T \left(\frac{\partial V}{\partial T}\right)_P$$

$$= \frac{\alpha_P}{\kappa_T} \left(\frac{\partial T_c}{\partial P}\right)_T$$

where $\kappa_T = -\frac{1}{V}\left(\frac{\partial V}{\partial P}\right)_T$ is the isothermal compressibility.

Now we are ready to evaluate

$$\left(\frac{\partial \ln \tau}{\partial T}\right)_V = \frac{1}{V} \left\{ \frac{3}{2} \frac{T_c^{1/2} \frac{\alpha_P}{\kappa_T} \left(\frac{\partial T_c}{\partial P}\right)_T}{|T - T_c|^{3/2}} - \frac{3}{2} \frac{T_c^{3/2}}{|T - T_c|^{5/2}} \right.$$

$$\left. \times \left\{ 1 - \frac{\alpha_P}{\kappa_T}\left(\frac{\partial T_c}{\partial P}\right)_T \right\} \right\}$$

$$= \frac{1}{V} \frac{3}{2} \frac{\alpha_P}{\kappa_T}\left(\frac{\partial T_c}{\partial P}\right)_T \frac{1}{|T - T_c|^{3/2}} \left[\frac{T_c}{T - T_c} + 1\right]$$

$$- \frac{3}{2} \frac{T_c}{|T - T_c|^{5/2}}$$

$$= \frac{1}{V} \frac{T_c^{3/2}}{|T - T_c|^{5/2}} \left[\frac{T}{T_c}\frac{\alpha_P}{\kappa_T}\left(\frac{\partial T_c}{\partial P}\right)_T - 1\right] \quad (16.18)$$

Finally, combining Eqs. (16.17) and (16.18) we have derived the E_V/E_P ratio as

$$\frac{E_V}{E_P} = \frac{1 - \frac{\alpha_P}{\kappa_T}\frac{T}{T_c}\frac{\partial T_c}{\partial P}}{1 + \frac{2}{3}\alpha_P\left(T - T_c\right)} \quad (16.19)$$

and this ratio can increase, decrease or remain the same as a function of temperature. This is the Bendler–Fontanella–Shlesinger equation. It is the dynamic nature of the DAD model and its combination of kinetics and thermodynamics provides this breadth of behavior (which is not found in simpler static free volume models). The E_v/E_p ratio can be further explored by writing T_c as

$$T_c = \frac{z|\Delta H|}{4k_B} = \frac{z|\Delta E + P\Delta V|}{4k_B},$$

$$\left(\frac{\partial T_c}{\partial P}\right)_{|T|} \approx \frac{z|\Delta V|}{4k_B} \qquad (16.20)$$

where T_c is identified with the critical demixing temperature associated with a simple Bragg–Williams (defect) phase separation transition (for nearest–neighbor pair interactions with equal occupancy of A and B sites), z is the lattice coordination number, k_B is Boltzmann's constant, and $\Delta H = \Delta E + P\Delta V$ is the decrease in enthalpy resulting from the formation of a defect pair with ΔE being the decrease in pair energy and ΔV being the decrease in volume. Substituting Eq. (16.20) for T_c and $\left(\frac{\partial T_c}{\partial P}\right)_{|T|}$ into eqn. (16.19) yields another form for the Bendler–Fontanella–Shlesinger equation

$$\frac{E_V}{E_P} = \frac{1 - \frac{z}{4k_B}\frac{\alpha_P}{\kappa_T}T\Delta V}{1 + \frac{2}{3}\alpha_P(T - T_c)} = \frac{1 - \frac{\alpha_P}{\kappa_T}T\frac{\Delta V}{\Delta H}}{1 + \frac{2}{3}\alpha_P(T - T_c)} \qquad (16.21)$$

16.8. The Prigogine–Defay Ratio

The Prigogine–Defay ratio

$$\Pi = \frac{\Delta C_P \kappa}{TV\alpha_P^2}$$

can be rewritten using the Ehrenfest relation

$$\frac{dT_g}{dP} = \frac{TV\alpha_P}{\Delta C_P}$$

as

$$\Pi = \frac{\kappa}{\alpha_P} \frac{1}{dT_g/dP}$$

If we replace dT_g/dP by dT_c/dP, then we can relate the Prigogine–Defay ratio to E_v/E_P as

$$\frac{E_V}{E_P} = \frac{1 - \frac{\alpha_P}{\kappa_T} \frac{T}{T_c} \frac{\partial T_c}{\partial P}}{1 + \frac{2}{3}\alpha_P(T - T_c)} = \frac{1 - \frac{1}{\Pi} \frac{T}{T_c}}{1 + \frac{2}{3}\alpha_P(T - T_c)} \quad \text{or}$$

$$\Pi = \frac{T_c}{T} \frac{1}{\left(1 - \frac{E_V}{E_P}\left[1 + \frac{2}{3}\alpha_P(T - Tc)\right]\right)}$$

There is interest in finding materials with $\Pi \geq 1$.

CHAPTER 17

Deterministic Random Walks

17.1. Bouncing Ball Through a Corridor and the Cauchy Distribution

The processes described in this chapter are not random walks, but in a deterministic process, averaging over a distribution of initial conditions generates a distribution of outcomes. Picking an initial condition at random generates an outcome at random and that distribution can by analogy be compared to random walk properties. Let's start with a simple model of a ball bouncing down a 2D corridor of length L and width d entering at angle θ.

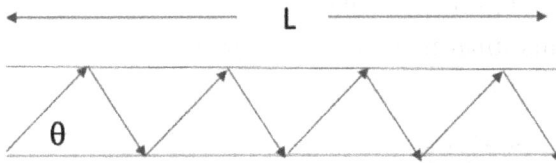

In each bounce the ball will travel a horizontal distance $x(\theta) = d/\tan(\theta)$. There will be $n = \frac{L}{x} = \left(\frac{L}{d}\right)\tan(\theta)$ bounces for the ball to exit the corridor. For a uniform distribution of angles, $f(\theta) = \pi/2$ this generates a distribution p of bounces from the distribution of initial angles

$$p(n)dn = f(\theta)d\theta = \frac{\pi}{2}d\theta$$

and $\frac{dn}{d\theta} = \frac{L}{d}\frac{1}{\cos^2(\theta)}$. Using

$$n^2 = \left(\frac{L}{d}\right)^2 \tan^2(\theta) = \left(\frac{L}{d}\right)^2 \frac{1 - \cos^2(\theta)}{\cos^2(\theta)}$$

Solving for $\cos^2(\theta)$ in terms of n, the distribution $p(n)$ for the number of bounces (jumps) to move a distance L,

$$p(n) = \frac{\pi}{2}\frac{d\theta}{dn} = \frac{\pi}{2}\frac{L/d}{n^2 + (L/d)^2}$$

where averaging over initial conditions produces the Cauchy distribution with an infinite second moment. This long tail captures the case when the ball enters the corridor with near to the vertical trajectory and very little horizontal motion.

In his *Physics Today* August 1999 article, George Zaslavsky showed that a particle scattering among a 2D array of flattened ovals produces a Lévy flight type of trajectory. The trajectory can take many bounces between two flattened parts of a lower and upper oval and then take a long trajectory down the spacing between ovals until it contacts a new oval.

17.2. The Geisel Map: Iterating Through a Curved Corridor

Another example is what if the wall is curved and perhaps in a curve that does not possess a Taylor expansion. Geisel and

Thomae studied this type of behavior with a mapping,

$$x_{t+1} = x_t + ax^z$$

It is perhaps easiest to see the corresponding behavior to number of bounces by going to the continuum limit,

$$\frac{dx}{dt} = ax^z$$

and integrating for a time T from an initial position x to final position L to arrive at

$$T(x) = \frac{(x^{1-z} - L^{1-z})}{a(1-z)}$$

a uniform distribution $f(x) = 1/(L - x_0)$ of starting points generates a distribution of escape times

$$\psi(T)dT = f(x)dx$$

where

$$x(T) = (a(1-z)T + L^{1-z})^{\frac{1}{1-z}}$$

with asymptotically x going as $T^{\frac{1}{1-z}}$ and $\frac{dx}{dT} \sim T^{\frac{z}{1-z}}$, thus

$$\psi(T) \sim \frac{1}{T^{\frac{z}{z-1}}} = \frac{1}{T^{1+\frac{1}{z-1}}}$$

for $z = 2$, we asymptotically arrive at the Cauchy distribution. When $\frac{1}{z-1} < 1$ the mean escape time is infinite.

17.3. Kicked Rotators and Oscillators

In the Geisel–Thomae map a z exponent is introduced and it appears in the escape time distribution exponent. The so-called standard map of a kicked rotator,

$$x_{n+1} = x_n + K\sin(2\pi\theta_n)$$
$$\theta_{n+1} = \theta_n + x_{n+1}$$

generates a mixed phase (x, θ) space with infinitely nested periodic orbits embedded in a chaotic sea. The phase space structure is sensitive to the value of K. The only nonlinearity is the
sine and no specific exponent is in sight, but different exponents
appear for different K values when generating a histogram of,
say, escape times from a region for an ensemble of initial conditions. The histogram can be treated as a probability distribution. In a *Physics Today* February 1996 article Yossi Klafter,
Gert Zumofen and myself show for $K = 1.1$ a period 3 orbit
upon magnification would be found to be composed of a period
7 orbit which upon further magnification would be comprised of
period 10 orbit, and so on generating what is called a cantori
structure. For $K = 1.03$ we calculated the long time tail distributions for leaving (being stuck near) period 3 and 5 orbits to
asymptotical have the forms $t^{-2.2}$ and $t^{-2.8}$.

Another example of this kind is Zaslavsky's kicked oscillator
model of a kicked electron in a magnetic field. In discrete
mapping form,

$$u_{n+1} = (u_n + K\sin(v_n))\cos\left(\frac{2\pi}{q}\right) + v_n\sin\left(\frac{2\pi}{q}\right)$$
$$v_{n+1} = -(u_n + K\sin(v_n))\sin\left(\frac{2\pi}{q}\right) + v_n\cos\left(\frac{2\pi}{q}\right)$$

The number q sets the symmetry and the orbits trace out
fractal patterns in (u, v) phase space. Like the standard map
only trigonometric nonlinearities are involved and long time tail
probability distributions are involved as in the standard map.

The picture on the first page is from an ensemble of initial conditions in a six-fold symmetric Zaslavsky map generating the star like picture. Single trajectories have Lévy flight type structure and transport probabilities will have fractional exponents. There are many open avenues of research for probability question for nonlinear maps.

These nonlinear maps are from the field of nonlinear dynamics which was the field that introduced the concept of chaos. Here we injected the notion of a probability distribution into a dynamical process by studying the behavior of an ensemble of initial conditions. George Zaslavsky's *Hamiltonian Chaos & Fractional Dynamics* (Oxford University Press, 2005) devotes a whole book to this topic.

A Few More Words

This work has given me the opportunity to acknowledge my mentors Elliott Montroll and Harvey Scher. Even though they are well cited, perhaps this will bring their accomplishments to a newer and wider audience and to show what grew out of a single 1965 paper, the CTRW paper. The autobiographical nature of this book, recounting personalities and travels, was only meant to show what a great time one can have as a scientist. This is also an opportunity to thank my random walk colleagues and co-authors, John Bendler, John Fontanella, Barry Hughes, Yossi Klafter, George Weiss, George Zaslavsky and Gert Zumofen. Great colleagues make life and work immeasurably enjoyable. Some go to conferences and make enemies, making friends is better. I have not mentioned other works and colleagues on different topics, but perhaps in a later essay if the response to this book is positive. Spending most of my career outside of universities, I never had students so this book gave me the opportunity to pass on some knowledge and history that I would have liked to teach. For students, this book will give you a basis for some understanding of some probability topics and mathematical techniques. In a sense this book is for beginners because the field of anomalous diffusion has grown enormously, at least two generations passed me. There are many avenues of discovery left and this is a welcoming community. Enjoy!

Index

www.ingramcontent.com/pod-product-compliance
Lightning Source LLC
Chambersburg PA
CBHW050602190326
41458CB00007B/2144

9789811232800